WONDROUS TRUTHS

WONDROUS TRUTHS

The Improbable Triumph of Modern Science

J.D. TROUT

OXFORD
UNIVERSITY PRESS

Oxford University Press is a department of the University of Oxford. It furthers the University's objective of excellence in research, scholarship, and education by publishing worldwide.Oxford is a registered trade mark of Oxford University Press in the UK and certain other countries.

Published in the United States of America by Oxford University Press
198 Madison Avenue, New York, NY 10016, United States of America.

Library of Congress Cataloging-in-Publication Data
Names: Trout, J. D.
Title: Wondrous truths : the improbable triumph of modern science / J.D. Trout.
Description: New York, NY : Oxford University Press, 2016. | Includes bibliographical references and index.
Identifiers: LCCN 2015030988 | ISBN 978-0-19-938507-2 (hardcover : alk. paper)
Subjects: LCSH: Science—History.
Classification: LCC Q125 .T74 2016 | DDC 509—dc23 LC record available at http://lccn.loc.gov/2015030988

9 8 7 6 5 4 3 2 1
Printed by Sheridan, USA

To Pat and Don Nadler,
for giving our children their good name,
and for so much else

CONTENTS

PREFACE

In the summer of 1999, I lucked into giving a lecture at the University of the Basque Country, in San Sebastian, Spain. While relaxing there, I wrote an article called "Scientific Explanation and the Sense of Understanding." I argued that philosophers may find value in constructing theories or models of explanation, but scientists are not much guided by these "requirements" on explanation. Instead, they accept explanations discussed at lab and professional meetings, explanations that "make sense" to their educated ear, explanations that convey a sense of understanding. This sense of understanding is not the exclusive product of those with true theories; it is equally enjoyed by people holding wildly false theories, like astrology or Creationism. And yet, I am a scientific realist. How can I hold that so epistemically feeble a notion as the sense of understanding could be part of the best explanation for the success of science? Referees and critics often wondered the same, and they were not as indulgent or comforting as the beaches of Spain. This book continues that story, on the epistemic role that explanation and the sense of understanding appear to have in the theoretical progress of science.

All told, this book attempts more than that story—a trifecta, in fact: a distinctive position in the philosophy of science, with sweeping themes and dramatic twists, all written for a popular audience. For people interested in the grand vision of science, it is natural that the history of science be treated as evidence for big hypotheses about the nature and structure of science. So philosophers of science of my generation either read works by or about historical figures like Copernicus, Galileo, Newton, Darwin, Maxwell, or Einstein, or surveys of the history of science. Often they read both. Books tangling with big themes of scientific progress and the very nature of science filled bookshelves. Books by Toulmin, Crombie, Grant, Lloyd, Lindberg, and Gillespie were all staples for philosophers of science using the history of science as evidence. But for sheer drama and controversy over big themes with wide reach, few books could match Popper's *Conjectures and Refutations*, Kuhn's *The Structure of Scientific Revolutions*, and Feyerabend's *Against Method*. These books advanced conceptual tools with which to understand the history of science. And, philosophers of science could use concepts like Popper's falsifiability, Kuhn's paradigms, and Feyerabend's flouting of method, to examine the way that evidence from the history of science might support various metaphysical or epistemological interpretations of scientific theories. We could also use those tools to assess the quality of the social sciences and challenge conceptions of scientific rationality.

Sleek, unifying stories about the growth of science attracted an educated public of nonphilosophers. As a new generation of philosophers of science found, these views were suggestive, even powerful. But they were also fraught. Armies of scholars unselfishly authored hundreds of articles and books trying to advance our textual understanding of canonical works by Popper, Kuhn, and Feyerabend, for example. But disputes over alleged exegetical inaccuracy will outlive us all. At the risk of boorishness, then, I will step beyond these disputes, framing issues in ways that don't depend on branded interpretations of a philosopher's view but on durable issues that concern the community.

Once we sidestep the sorts of narrow issues that obsess us in graduate school (like the appropriate brand of realism, or the best account of reference to capture theoretical continuity), it is easier to keep the big issues of scientific progress in view. The technical problems can be interesting, but they are issues for another day.

My decision not to engage these disputes is not a mark of disrespect. Rather, I have tried to protect important topics that are fresh and promising by not scattering them among legacy scholarship in the philosophy of science. So, while I take up big questions like "What explains the steep rise of science since Newton?" and "What overall view of the history of science must a person have to explain this advance?," I have steadfastly resisted getting distracted into scholarship. As a result, I was able to write a relatively short book with grand themes and, I hope, practical content.

Behind every achievement is a benefactor. This book has many. Two of the most influential are Richard Boyd and Frank Keil. Some of the first conversations I had with Dick concerned radical epistemic contingency and the rise of corpuscularism, and with Frank about the conceptual and categorical basis of explanation.

While explanation is a canonical topic in the philosophy of science, research on explanation hasn't stood still. The last 50 years have seen ingenious work on explanation, by Hempel, Salmon, Kitcher, Woodward, Thagard, Strevens, and many others. Joining this research on explanation more recently is an ever-growing band of psychologists, including Tania Lombrozo, Ben Rottman, Philip Fernbach, Deena Skolnick-Weisberg, and Andrew Shtulman, to name just a few. And most of them are early in their careers, so this is just the beginning.

Many philosophers, psychologists, and friends have made this book a fun and relaxing enterprise. For useful comments, conversations, or inspirations concerning the topics of this book, I would like to thank Paul Abela, Tania Lombrozo, Deena Skolnick-Weisberg, Stephen Grimm, Erik Angner, Carlo Martini, Susanna Siegel and Jason Rheins. My research assistants over the years, Matt Kelsey, Corbin Casarez, Jay Carlson, and Marcella Russo, did much to improve this book. Two others, Andy Kondrat and Stephanie Hare, made substantial suggestions and labored patiently over the organization and prose. Maya Mathur, David Reichling, Vandana Mathur, Norma Reichling, Vincent Hausmann, Peter Sanchez, and Kathleen Adams are all friends who, almost thoughtlessly, deliver clever insights in casual conversation. I have used those insights equally casually. I also want to thank my invaluable editors at Oxford, Emily Sacharin and Peter Ohlin, the expert copy editor and manuscript shepherd Maya Bringe, and Loyola University for a leave during which I could write this book.

At just the right time, John Norton worked out a stint for me as a Visiting Scholar at the Center for the Philosophy of Science at the University of Pittsburgh during the fall of

2010. The Center is a wonderful place. I had a great time talking and listening to faculty and visitors alike: Edouard Machery, Richard Samuels, P.D. Magnus, Kareem Khalifa, James Woodward, and Heather Douglas, to name a few. It was there that I presented material from Chapter 2 of this book and tested out ideas presented elsewhere in this work.

I owe a special thanks to three outstanding philosophers—Michael Bishop, Joe Mendola, and Michael Strevens—who read the final draft of the manuscript and gave me substantial and detailed comments. I am lucky to be in their orbit.

Over the last few years or so, I have presented this work on psychological fluency and the sense of understanding, historical contingency, scientific realism, and ontic explanation, at conferences in the United States and abroad. I have addressed these issues to philosophers and historians of science, cognitive psychologists, and neuroscientists at university campuses like University of Texas, Northwestern University, Indiana University, University of Pennsylvania, Washington University, University of California at Berkeley, and the University of Pittsburgh. I have tested out drafted material from this book at Washington University, at the Cognitive Science Society 2008 meeting (where I first presented the fluency account of the sense of understanding), at Indiana University, and at the University of Pittsburgh. I also presented a series of lectures at University of Helsinki's TINT Centre, a vibrant place teeming with young talent. I want to thank its Director, Uskali Mäki, and thank Carlo Martini for being such a gracious host. Carlo also kindly arranged an Author-Meets-Critics session for *Wondrous Truths*, and served as friendly critic along with Jaakko Kuorikoski and Caterina Marchionni. I am grateful for their comments on that occasion, and the book has benefited. I have also profited from feedback at the Midwest Epistemology Workshop at University of Missouri, Columbia, in 2015.

Finally, I am grateful for my family—my wife, Janice, and our children, Jack and Jessie. I am also grateful for a job that affords me the freedom to enjoy my family to the fullest.

J.D. Trout
October 2015

WONDROUS TRUTHS

1

WONDER AND THE FEELING OF UNDERSTANDING

A HISTORY OF WONDER

Anyone who has watched a spider spin a web—or watched a child watching it—will know the story I am about to tell about the beauty of wonder. Everyone who has stood in awe of the creations of nature or the richness of our inner experience—whether it be the Grand Canyon, meteor showers, a performance of "Nessun Dorma," subtle subatomic interactions, a death that is peaceful and happy, or an act that is utterly selfless—knows the majestic and solemn bond we feel to a world grander than could be conjured by our minds.[1]

Wonder can take any form. We can wonder at a legacy or story line that reaches back in history, richly connecting us to remote ages. When we look at the Laetoli footprints in a Pliocene bed of sandstone in Tanzania, we are observing the remnants of early human parents and children memorialized in the path. Their footprints, preserved in volcanic ash, proceed in a sequence and into the distance. They speak of family groups walking together. There are adult and child footprints side by side. We are connected by the familial bonds in those ancestral footsteps of Laetoli, and we use them to wonder about the story behind these steps. How did the mother show her affection or evade predators while out on this walk? How did the child misbehave? How long did they live? Are the two sets of footprints close enough that the mother and child might have been holding hands? Where were they going? Underlying this wonder is a feeling that we understand that same love the mother felt for her child.

We can wonder at the endless variety of life forms, elegantly adapted to the most inhospitable of environments. Take the alien appearance of Earth's deep-sea occupants. At depths that would crush a powerful human, the delicate "dumbo octopus" (*Grimpoteuthis*) lives two miles beneath the ocean's surface, doing a kind of haunting, solitary ballet. What is their closest relative? How does a gelatinous creature resist collapse at such high pressures? Do the "ears" (the flaps) have a function? Underneath this wonder is a sense of foreignness, an attempt to understand another being's world.

Humans have always had moments of wonder, of elevation, in the presence of the world's beauty. This feeling may manifest itself as a deep connection to humanity long passed or to a place we may never see. It may be a feeling of being lifted up, inspired by events and visions greater than oneself. This inspiration may spring from our sense of insignificant drift in a universe of unimaginable scale, or it may well up from the expressions of another primate, in whose face we recognize common lineage.

Aristotle said that philosophy begins in wonder. To Thomas Carlyle, it is not philosophy, but worship, that begins in wonder. To Samuel Johnson, wonder is the effect of novelty on ignorance. Each of these notions points to the same fundamental notion: in the classic balance of human talents and nature's secrets, human wonder arises from our inherent cognitive limitations driving a wedge between those talents and secrets. We would never feel wonder if we immediately understood every puzzle, problem, and proposition we faced.

It should not be surprising, then, that the same cognitive limitations that make wonder possible in turn make mental shortcuts necessary when we try to understand the things that make us wonder. Between wonderment and understanding are dozens of conceptual gaps that must be spanned in the geography of any theoretical outlook: star destruction by black holes, carbon dioxide emission from the earth's seas, solar neutrinos that flow through the earth, and bats that crave sugar to reduce ethanol toxicity. The complexity of these phenomena are untrackable without the prosthetics of models (however they may mislead) and simplifying assumptions (that skate over crucial details and qualifications). No matter what we learn in school, electricity is not, after all, a flowing liquid, and an atom is not a little solar system. These information compression techniques are necessary—and necessarily misleading—tools. But once we use these cognitive shortcuts, there we stand, viewing the grandeur of a beautifully complicated vista from a simplified sightline.

Wonder is so striking an experience, and so central a feature of the human effort to understand, that our favorite stories portray scientists in the grips of a childlike joy, or a sense of divine elevation or reverence, when they make a discovery. The history of science displays episodes such as Kekule's and Archimedes's famous "Eureka" moments, which prominently feature dramatic journeys from insight to compelling explanations.[2] These scenarios convey a sense of understanding, producing an explanation that "feels right." This psychological sense of understanding is a kind of confidence abetted by hindsight: it is intellectual satisfaction that a question has been adequately answered.[3] Sometimes we simply wonder without asking "why," as when we wonder at the majesty of a mountain vista, and let the austere beauty wash over us. But characteristically, the pleasant feeling of wonder is connected to the sense of understanding in a reward cycle we will examine in Chapter 3. The sense of satisfaction can be described simply as a heightened confidence that one enjoys an accurate description of the underlying causal factors—that one has hit upon the truth.

But it will be important to remember that confidence is, notoriously, not an indicator of truth. Elated scientists regularly announce their grand discoveries, only to find that their observations were unreliable, their calculations mistaken, or their background theories defective. This sad fact raises the most persistent puzzle about the boundless discoveries of our finite minds: how did modern science, probably the greatest of intellectual achievements, emerge from a psychology that can experience wonder without understanding, explanations without accuracy, and intellectual adventure without reliability?

Scientific Explanation and Manageable Wonder

Good explanations are accurate causal accounts of the things we wonder at, things we want to understand. The origin of explanation may be found in the earliest efforts to see the workings of other minds, to bring us information to manage family members and officiate social disputes. We explain our children's recalcitrance, a fiancé's recent ambivalence, or strife in the family and at work. But explanation's earthy origins haven't grounded it; we have used it to account for the grandest of mysteries of the natural universe. Why does lightning produce a thunderclap? Why do some cancers respond better to chemotherapy than others? Why is there a red shift in light coming from cosmologically remote objects? Why do some children with the same level of hearing loss respond differently to cochlear implants? Why does the FOXP2 gene regulate such a wide range of processes, from gut to speech development?

The history of science reads like a biography of explanation, so we should not be surprised that the push for good explanations elevated science from medieval alchemy to electrochemistry or from a Newtonian ballistic physics to quantum physics. And though the attempt to explain has existed as long as we have been able to wonder, if you look at a long science timeline from prehistory to the present, there is a steep curve of theoretical discovery that explodes around 1600, primarily in the West. This fact leads to more wonderment: Why the West? Why so quickly? This book answers these two questions by touring the findings of neuroscience, psychology, history, and policy. And the answers are surprising. The central idea of this book is that science in selected areas of Europe rose above all other regions of the globe because it hit upon successive theories that were approximately true through an awkward assortment of accident and luck, geography, and personal idiosyncrasy. Of course, intellectual ingenuity partially accounts for this persistent drive forward. But so too does the persistence of the objects of wonder.

When we talk about explanations that are approximately true, we acknowledge that explanations can be true or false, better or worse, when giving an account of unobservable causal mechanisms. A good explanation is a true one, but unfortunately, there is no psychological homing system for the truth. We accept explanations because they *feel* right and they produce that sense of understanding. But historically, false explanations have felt about as right as true ones. If scientific progress depends on having good explanations, how can you calibrate that feeling so that you only get that feeling about good theories? The answer is disarmingly simple but maddeningly evasive: that feeling becomes more reliable the more you are surrounded by good theories. Once you have a good theory, some of the next ones will be better than others, but all of the plausible alternatives will be better than you ever had before.

This brings us to another maddening fact: good explanations remain good regardless of whether or not they suit the purposes of *human* explanation, and they are good independent of whether anyone knows or understands them. This may sound strange at first, but that is because we routinely confuse two different things: providing a good explanation and framing or elucidating an explanation that is good. It becomes more palatable once you consider explanations of natural events. For example, the honeybee's failure to find nectar high above the ground is explained by the fact that altitude information is not conveyed by the honeybee's vibratory dance. This is a true explanation, whether or not the honeybee understands it or can ever understand it. Or consider a good explanation for thunder: lightning heats a column of air, causing the acoustic event from heat-expanded molecules. The five-year-old child might instead explain thunder by saying that the angels are bowling. However reassuring to a child, this is an objectively bad explanation; it fails to accurately describe the actual causes that produce the thunder. And the molecule expansion explanation is a good one no matter what the audience's age, even if people only come to appreciate it into adulthood. So it doesn't add much to say that it is an explanation *for us*, beyond simply saying that we are smarter than honeybees and children.

If we know that honeybees and children don't have the right explanation for certain phenomena, then how will we know that *any* explanation is good, let alone true? Isn't it possible that many explanations are beyond intelligent adults as well? There are, after all, reliable rules of inference that we use without knowing it, and without even being able to recognize the rules. So a method can be reliable without one knowing that it is reliable, just as a pitcher might throw a wicked curve without being able to say how he does it. That doesn't mean a generally reliable method is exceptionlessly good: whenever a method is generally reliable, it may occasionally produce a falsehood. But once we hit upon a good theory, a method's reliability may be like the man with no name—it comes out of nowhere, without a pedigree. For example, conceiving unobserved reality as a collection of particles has been popular at least since the time of Democritus. But the methods used in developing this atomistic conception never worked until we got it right about the specific properties of these corpuscles—until we assumed correctly that they were round and not pointed, or that they were prone to agitation by heating. Now, it might be nice to be able to say what facts make a theory true, but this isn't necessary for such explanations to be good. Tacitly assuming the truth of something that in fact turns out to be accurate may be enough for a successful practice or for a good explanation.

This approach to explanation is called "ontic," in which the quality of an explanation is determined by its possession of certain objective factors, like its accurate description of causal factors. One consequence of this ontic account is that an explanation can be good even if no one is in a suitable position to comprehend it. By acknowledging these natural contours and limitations of human understanding, the ontic view allows us to reassess our intellectual trajectory since the Enlightenment.

Not everyone accepts the ontic view of explanations. There are Protagorean theories of explanation as well, which treat an explanation's ability to satisfy human cognitive purposes as the measure of its goodness.[4] As a result, for an explanation to be good, it must be an explanation *for* someone; someone must understand it. And understanding it will require that it is couched in a vocabulary and theoretical outlook that illuminates our existing theoretical knowledge. In short, an explanation's goodness is entirely dependent on how well it suits the human purposes of explanation. But in the end, the Protagorean view of explanation is too modest to account for the striking theoretical and practical achievements of science. Good explanations aren't just stories; they are *true* stories.

How did we become familiar with, and fluent in, the basic scientific explanations of the West? How did we latch on to stories that are *true*? We achieved this fluency, this feeling of easy processing, from exposure to the theoretical tradition in physics and chemistry associated with Robert Boyle and Isaac Newton in the last half of the 17th century. In the last 400 years, this "Newtonian," corpuscular tradition in physics and chemistry dominated natural inquiry. But it wasn't always that way. Before that, scientific development in Western Europe and the Americas was dormant for two millennia of recorded history, while much of the rest of the world created ingenious methods for controlling their environments. By 500 BCE, China, India, the Middle East, and northern Africa had all constructed wells and irrigation, grand buildings, and bridges. They bred livestock and produced accurate calendars. They had astronomical theories and practical medicine, and keeping this all afloat were distinguished traditions in mathematics. And long before the science of Germany, France, and England took off, the Islamic world from the Arab Peninsula to Cordoba invented and perfected algebra, correctly estimated the curvature of the earth, and had experimentally demonstrated both the rectilinear propagation of light and a palliative for constipation. Their predictions were useful and more accurate than you would expect by chance.

But useful technology does not necessarily depend on a true theory, and by 1700, India, China, Japan, and the Islamic world had only their practical expertise. What stood between the continuous and vaulting science of 17th-century London, Paris, and Florence and the disparate technologies of other heavily populated regions both within and beyond Europe? In short, an approximately true theory, one that accurately explains the causes of our objects of explanation. When you luck into a true theory, your science can be off and running.

Fortifying the common view that science is an accumulation of detailed observations, one textbook describes Copernican theory as an inductive tool, documenting "observations leading to a general statement or natural law."[5] On the contrary, it wasn't the power of experimental method that drove science forward in the 17th century; a sound experimental method is no substitute for a good theory. That is, the Experimental "Scientific" Method did not cause modern science. If that were all it took, we would have had modern science under Grosseteste—the 13th-century English statesman and scholastic philosopher who set out the rudiments of experimental method over four centuries before Newton. Or we would have had modern science under medieval Islam, in which some of the same rules of experimentation pre-dated Francis Bacon by at least four centuries.

History, as well, has romanticized Newton's brilliance and supposed dogmatic approach to the scientific method in jumping from flawed alchemy to the beauty of modern physics, in a tradition captured in a line from a poem by Alexander Pope: "Nature and nature's laws lay hid in night. But God said 'Let Newton be,' and there was light."[6] But these treatments always overemphasize the role of the scientist, leaving aside ingredients like resources, instrumentation, patronage, and the best insights from otherwise poor theories that fortuitously align. Contingency and a series of lucky events can explain the rise of modern science far better than any adherence to dogged application of the Scientific Method.

If people find implausible the story of the big leap—from alchemy to Newtonian physics—perhaps it is the "representativeness heuristic" leading them astray.[7] They just can't believe that something true came from something so false. Of course, this contrast is unduly dramatic. Newtonianism was not The Final Theory, and alchemy was not wholeheartedly absorbed by the new physics. But Newton's theory allowed enough of the hidden structure of matter to be tested and revealed to sustain a research program that unified nearly every corner of physics and chemistry for the next 200 years. The leap was not just big but fast.

Furthermore, the success of modern science isn't a single, irreducible fact. A full explanation must appeal to many facts at once: there is the correct corpuscular hunch in Boyle's period, the rise of scientific academies, the ever-growing circle of scientists and communicative network of investigators, the steep increase in steady and reliable science funding, and a host of other causes.

With this tailwind, the Newtonian hunch spread mature science to every populated continent and promptly steered the course of human civilization there. But we understand very little about the nature of its growth. In trying to provide an answer for this rapid expansion, most often we get a tired appeal to the experimental method and the role of reason. But my answers are much different: the lesson of the fast spread of 17th-century science is the remarkable power of psychological fluency (coming up in Chapter 2) and the remarkable power of accident (described as "contingency" in Chapter 5) once we landed on beliefs about the unobservable world that are roughly true.

After 300 years of trying to explain why the science of the West emerged successful and still dominates the world, we do not have a very good explanation. In addition to the myth we've already been told about the application of the Experimental Method, there is an endless variety of stabs: we were chosen by a Christian God, we had a muscle-bound work ethic, and we had the accumulated capital to subsidize innovation. But none of these explanations convinces. To take them in quick turn: many Christian nations languished, many worked hard but abortively, and many relatively wealthy societies with patronized science stalled for centuries with an unproductive theoretical science. Curiosity, drive, and sense of wonder only take us so far; at some point, the world has to start cooperating with our vision.

My explanation for this result is at once classic and controversial: you can think of it like an actual take-off. ***Science took off as we started to hit upon theories that got it right enough about the unobservable world***—right enough to start rolling toward takeoff. You don't get off the ground if the upward force is almost or just as great as the downward force of gravity, and theories that fell far short of that spun their wheels on the runway. It doesn't matter whether Boyle and his immediate predecessors saw in late alchemy the promise of a new physics. All that matters is that the roundness, hardness, and elasticity of alchemy's corpuscles were recruited for the new physics. These are the properties that in fact regulated Boyle's and Newton's beliefs and led to the decisive command of this theory over the growth of science in the populated centers of education in Europe.

So why is it so difficult for people to explain the success of science? The main reason is that the success of science rests on getting things right about objects that are difficult or impossible to observe with the unaided senses. Whenever an object of explanation is not open to casual inspection, the explanation is more prone to be disputed. And it is a sense of psychological fluency that makes something seem open to casual inspection. It is the kind of fluency you feel when you are asked to compute 7 + 4 = 11 but not when you are asked to perform the same computation in base two, the fluency a right-hander feels when asked to play catch but not when asked to switch to lefty, or the fluency you feel when you're asked whether it is true that birds of a feather flock together but not when you are asked whether birds of a feather flock conjointly. Sometimes this fluency emerges either spontaneously from a new solution to an old problem (like Kekule's discovery of the ringlike structure of benzene), or from a routine articulation of existing concepts to solve old problems (like Newton's Law of Universal Gravitation to address Galileo's puzzles over falling objects) or new problems (like Boyle's Ideal Gas Law). Even the most jarringly novel insights that interrupt the fluency of outmoded concepts soon become fluent themselves when people come to see how the novel proposal makes sense of a wider range of phenomena.

Psychological fluency comes prepackaged with a unique kind of feeling, a sense of fluency that provides a fit with the facts, marked by an ease of processing. To the knowing subject, navigating by this sense of understanding feels like tracking true north.

But as we have already seen, this keen sense of fluency unfortunately is no compass. When you hit on an accurate theory, psychological fluency is a potent tool for discovery. When, on the other hand, your most familiar and beloved theories are lousy, this same psychological fluency leads you down countless blind alleys. And this asymmetry is what explains both why the history of science is a long period of stagnation amidst flurries of activity, and the rapid and sustained ascendancy of theoretical science in the modern period.

This book tells the novel and no doubt controversial story of scientific progress. It is a story of human biology, historical contingency, and routine hunches, all ultimately directed by accurate theories and true explanations. Like the progress of science itself, this book will wander down many different paths to bring together the whole story. This first chapter has set the pins in place; the remaining six chapters will hopefully knock them down. Chapter 2 will show how we can think we understand an explanation even when we're so very far from the truth. We use mental fluency, the ease with which we process ideas, as a cue that we understand. Unfortunately, however, that cue is far from reliable: it is not necessarily the case that a fluent explanation is attached to the truth. We are led astray by our errant Sense of Understanding, and scientists are no more immune to this phenomenon than the common person. The history of science is a story of false explanations that people "knew" to be true. So we have to ask, how *can* we know when our explanation, our scientific theory, is true? Chapter 3 delves deeper into how we explain and understand, showing learning and understanding as existing as part of the physiological reward cycle: it feels good to understand. Unfortunately, when it is difficult to trace the source of our success and failure, that good feeling arises when we are wrong just as often as when we are right.

Chapter 4 will introduce the inductive principle of Inference to the Best Explanation as a garden-variety form of explanation in both everyday life and scientific pursuits. A simple example widely used: there is some class of observed phenomena, like droppings on the cupboard shelf. Together with the sounds behind the baseboard, the best explanation for these phenomena is a (currently unobserved) mouse. Put another way: a mouse behind the baseboard, if present, would provide the best explanation for the droppings we see and sounds we hear. We then infer that there is a mouse behind the baseboard. Could we be wrong? Sure. But the inductive case for a mouse is very good. Scientific explanation, as it turns out, uses the same form of reasoning to infer everything from evolution to heliocentrism.

Of course, as noted, the history of science is littered with poor explanations. How, then, did science ever progress, and how did we get to the good explanations? The answer lies in Chapter 5. Science progressed because the theories being used were approximately true. And these theories didn't arrive through meticulous application of the Scientific Method; they arose through accident and contingency. An interesting side effect occurs when we happen upon theories that are approximately true, which ties into the ontic account of explanation: we don't have to understand the theories we use for them to be true. A good explanation is an accurate one, regardless of our understanding of it. Just as the spider doesn't have to understand how it spins its web for it to be an effective fly catcher, we don't have to understand how twin particles are connected across the universe for it to be true and for us to use that knowledge.

Chapter 6 tells the story of that great leap in scientific progress: Newton's guess about the nature of corpuscles happened to approximately track the truth of the universe. His hunch was able to best explain the mechanisms of the world and as such was useful. The great leap forward in science was not a necessary event that had to occur when smart people ground out the truth; it was a contingent happening, based on accidents of history, geography, and just plain luck. Finally, in the concluding Chapter 7, we will look prospectively: what does the story I've told in these pages mean for our future?

Knowledge begins with wonder and leads to explanations of how the universe works. But because the same cognitive limitations that lead to wonder also so easily lead us astray as we try to track the truth, along the way to good explanations we've been helped by contingency and simple luck. This book tells the story of both those limitations and our success in overcoming them.

2

EXPLANATORY FLUENCY AND THE FALSE CLIMB

THE HISTORY OF SCIENCE TRACES an uneven but unmistakable ascent, and a substantial proportion of this theoretical progress depends on accepting explanations that are ever more accurate. Kepler's explanation of the duration of orbits in terms of elliptical orbits; Boyle's explanation for the expansion of gases under heat in terms of the corpuscular composition of matter; and the explanations of Semmelweis, Snow, and Pasteur for the spread of disease in terms of microbial transmission, all described causal mechanisms and relations that oriented future research toward issues that could then be handled with much greater accuracy.

Because these causes are nearly always unobservable and typically complicated, many people, even trained scientists, rely on simple rules or heuristics to track and measure their properties. Weighing an explanation is an informal process, and one heuristic we often use is to assume an explanation's accuracy if it supplies a psychological cue: a sense of understanding. As a psychological matter of fact, we are led to accept explanations, scientific and otherwise, by the sense of understanding they convey. This simple view assumes an unadorned conception of explanation: an explanation is the description of underlying causal factors that bring about an effect.[1] We accept an explanation because it "feels right": it offers a phenomenologically familiar *sense of understanding*.[2] But clearly this sense of understanding is not always a reliable guide to truth (surely you can think of times you were certain you "got it," only to find you were woefully misinformed), nor a necessary or sufficient condition for good explanation. As Paul Humphreys, a philosopher of explanation, puts it: "It is no explanation to provide a distorted representation of the world, and the 'understanding' induced by such incorrect models is illusory at best."[3]

I will argue that even though the sense of understanding is an unreliable cue, it is widely and heavily used by scientists to justify their acceptance of an explanation.[4] In proposing their completed DNA model—which provided the explanation for the genetic basis of heritable characteristics—Watson and Crick expressed the power of the intellectual pleasure and the overwhelming feeling of psychological fluency it created: "[A] structure this pretty just had to exist."[5] Cosmologist Steven Weinberg hints at the hedonic basis of explanatory impulse: "Scientific explanation is a mode of behavior that gives us pleasure, like love or art."[6] When we think we understand something, we feel good about it, and thus we accept it.

Part of this good feeling found in explanations is that we are often driven by our intuitive assessments about the quality of the explanations we already possess. We stop searching for explanations for a number of reasons, but surely one is that we have achieved an adequate or "working understanding" of some object. As the psychologist Jonathan Haidt puts it:

> Once people find supporting evidence, even a single piece of bad evidence, they often stop the search, since they have a

> "makes-sense epistemology"[7] in which the goal of thinking is not to reach the most accurate conclusion; it is to find the first conclusion that hangs together well and that fits with one's important prior beliefs.[8]

This fit with prior beliefs is what many epistemologists, too, describe as a central feature of belief acceptance, and the chief evidence of the power of the sense of understanding: "That belief has a certain felt attractiveness or naturalness, a sort of perceived fittingness; it feels like the *right* belief in those circumstances."[9]

Described in this way, as a feeling of confidence or correctness, the sense of understanding is immediately recognizable. But we come to that feeling in many different ways, so it has no single cause. Sense of understanding can come from a variety of human biases, including overconfidence, hindsight, and confirmation. And a sense of understanding can relate to different psychological destinations. So when psychologists named the many senses of understanding, they brewed a veritable alphabet soup. They have measured FOK (Feeling of Knowing), IOED (Illusion of Explanatory Depth), FOR (Feeling of Rightness), and JOL (Judgments of Learning). These feelings are all tributaries in the torrential flow of our experience. Taken together, they are an impure and awkward lumping of cues. But this should not be terribly surprising. People embrace beliefs of all sorts simply because they offer a feeling of understanding, of fit, with other beliefs we hold. So we believe that not dressing warmly in cold weather brings on colds, that shaving hair makes it grow in thicker, and that hot foods cause ulcers. In each case, there is a similarity between the purported cause and effect that gives the believer an excuse to entertain the thought: being cold gives you an illness that makes you shiver; shaving transforms soft, thin hair into stiff (and thus thick) stubble; and hot food causes stomach inflammation.

The problem is, we are *generally* suboptimal as reasoners, sometimes severely so, and that fault, apparent in many different fields of study, carries over into our sense of understanding. In these fields, we don't muzzle our confidence, our hindsight, or our confirmation bias when we reach beyond simple tasks.

While philosophers and scientists sometimes exaggerate the logical rigor or formal purity of explanations with talk of deduction, laws, and models, these expressions of wonder and clarity are in no way exotic. Together, they establish the important role of psychological factors in the *acceptance* of an explanation. We accept an explanation by experiencing and evaluating feelings like pleasure, accessibility, and confidence. *How* scientists use these feelings in explanation choice is an empirical question, and the attitudes already described provide some hints. On the other hand, *whether* scientists *ought* to be guided by these factors is a normative question. We will examine answers to the latter question by assessing the evidence of the reliability of these feelings as a cue of accurate explanations.

Put starkly, when it comes to mental processing, easier is better liked, and as a result, the sense of understanding is stronger. In science, explanatory prototypes that free up processing space in the brain are deemed more attractive, and more accurate, whether or not they actually are.

Still other features get exploited as our brain efficiently favors information we can process easily over information that draws more heavily on shrinking resources. Take vision, for example: contrast and clarity are two of the most visually primitive elements for the eye to pick out, and art scholars and commentators perennially report the contribution of contrast and clarity to the aesthetic appeal of the objects that have them.[10] From a psychological fluency approach, it is easy to see why. Good information is hard to come by, and the easy-to-process elements of contrast and clarity—like figure-ground contrast—make it easier to extract information from a scene.[11] Easier to process means better liked. And once again, the fluency caresses us like a warm bath. Even the simple circle appears prettier when presented with a figure-ground contrast that is friendly to fluency.[12] Virtually anything that leads to greater fluency also leads to the desired somatic marker, a more positive affective judgment. One might ask what the aesthetic pleasure of contrast or the enhanced credibility of legible writing has to do with theory acceptance in the history of science. While these effects in particular don't, people tend to trade fluency for theoretical continuity. This continuity with a familiar object—be it a figure, a written word, or a theory—produces a sense of understanding, and it is this feeling of understanding that in turn drives the acceptance of true and false explanations alike.

Or consider overconfidence, which seems to cast a spell more powerful than all the other cognitive biases. Why are we suckers for overconfidence and bluster? When we don't have much information about the accuracy of a person's judgment, we are very influenced by their level of confidence.[13] Overconfident people can often get by on bluster, despite the fact that bluster is unrelated to accuracy.[14] But why in the world are we more likely to believe them over someone with lower confidence (and whose honesty calls them to express a reasonable degree of uncertainty[15])? When people in the same group give advice, say, to policymakers, overconfident people have the persuasive edge[16] and are regarded by others as more knowledgeable.[17] So if you only have the kind of information you get from the pundits' roundtable, it is important that the person, the "expert," actually be accurate.

There are delicate psychological reasons that people rely more heavily on those who radiate confidence. People strive for greater certainty, for instance, and so wish to affiliate with those who seem to possess it already.[18] But there is a simpler explanation for our deferral to the chronically overconfident. We listen to abbreviated or complicated explanations all day and can seldom independently verify what people tell us. But without this authentication, we cannot adequately appraise their sources' reliability. In the absence of any evidence for ineptitude or deceit, we make the practical assumption that people are capable of estimating the quality of their own self-ascribed knowledge. Understandably, we are more likely to make a high-confidence assertion when we have better, more reliable information than a low-confidence assertion.[19] We have no real theory to rely on when determining whether people's confidence judgments are likely to be well grounded or not.

Familiarity makes for a fluent, good feeling in all of us, and this includes scientists as well. Of course, for every accurate explanation offered by Kepler, Boyle, and Semmelweis, there was a false one by Ptolemy, Geber, and Paracelsus that was prompted by a similarly seductive sense of understanding. Thus, fluency (and with it, the sense of understanding), given its attraction to false information, cannot by itself explain the ascent of science. Instead, it identifies a heuristic, albeit an unreliable one, used to pursue and accept explanations. To describe the persistent accuracy of modern scientific explanations, we must invoke the epistemic contingency of the history of science: when science advances, it is because scientists are drawing their explanations from theories that happen to be accurate. The most plausible psychological account of the psychology of explanation leads to the most dramatic story of scientific progress. To make this case, we need to explore a fluency heuristic that produces a pronounced sense of understanding that gives way to fantastic progress but can also lead to an equally decisive failure.

The fluency heuristic we will explore affects both the big and the small: fluency effects even reach into assignments of credibility to specific assertions. People found a product more innovative if its description was printed in a difficult-to-read font.[20] People were more prone to accept a claim, such as "Osorno is a city in Chile," when it was easier to read on a colored background.[21] And sayings that rhyme are judged more accurate than their nonrhyming equivalents ("What sobriety conceals, alcohol reveals" vs. "What sobriety conceals, alcohol unmasks"), even when people are asked to separate how it sounds from what it means.[22]

Why does ease of processing increase the expected probability of an explanation's truth? The most plausible account appeals to our cognitive boundedness—our mental limitations—and the indelicate rules of thumb that our need for simplicity demands. Without a link between fluency and truth, our limited reasoning capacities would face an unmanageable number of tasks. We can only keep a handful of information chunks in working memory, and even then for only a few seconds. Our lives would slow to an awkward struggle if we had to deliberate laboriously about every decision. It is no wonder, then, that we need some heuristic to streamline a clogged system. So the process goes something like this: it feels good to feel like we got it right, and we often get that feeling when we are. Indeed, we take cognitive shortcuts to generate this feeling of understanding. The product is a somatic marker, a good feeling, that matches the cognitive goal of easy processing. In short, this result feels right. Arriving at it was easy and satisfying, and I could perform the task while doing other things; I had cognitive energy to spare. If there were any real chance I was wrong about any outcome I could so completely attend to, I would have felt the cognitive effort or strain; I would have been aware of the snags. But it feels effortless and complication-free. In a nutshell, that is the fluency heuristic. So, just how reliable is this fluency cue?[23]

THE FALSE CLIMB: A SENSE OF ASCENT

Inexperienced pilots, and pilots flying without instruments, face a special peril at night: The False Climb. Disoriented in the dark, they blindly follow the feeling that they are gaining altitude. They even adjust their controls accordingly. But the pull they feel as though they are advancing upward against gravity is really the result of acceleration downward as they steer their plane, tragically, into the ground (Figure 2.1). This "false climb" is a perceptual illusion rather than a suicidal wish. Their bodies are telling them this is the right thing to do. How can this be? How can a feeling steer us so wrong? How can so primitive an intuition be so unreliable?

With all the normal ground cues shrouded in darkness, accelerating straight ahead, or even downward,

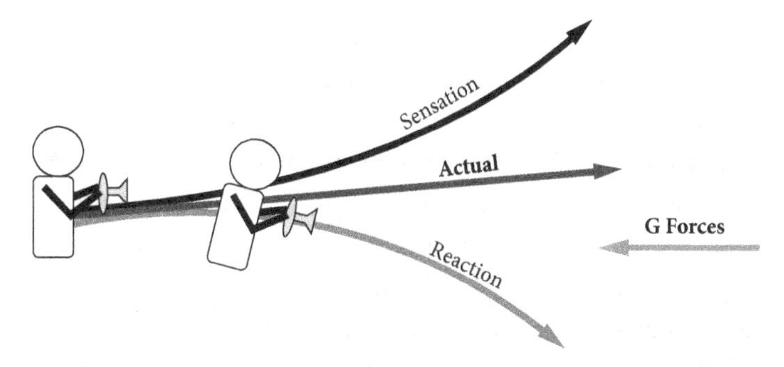

FIGURE 2.1 The forces producing a false climb experience, the fluent but misleading sensation delivered by the otoliths that you are gaining altitude when you are actually accelerating toward the ground. Often fatal for pilots and passengers, false climb experience usually occurs on dark nights when pilots do not smoothly transfer attention to their instruments.

feels as though we are climbing. This is because when the straight acceleration pushes back the otolith organs in our inner ear, our brain interprets this event as if our head is tilted backward, which is the normal condition when we have this experience. So it feels as though our head is tilted backward and we are climbing against gravity. But we only *think* we are climbing. Instead, we are accelerating toward the ground. This is because steeply accelerating travel is only about a century old. But the low-G world of our primitive ancestors is the world in which our otoliths evolved. In that environment, gravity was the main force that influenced our inner ear. Humans take signals from the otoliths as indicating the position of our head. Our inner ear only knows accelerative forces; it doesn't know up or down, so evolution has led us to think that any accelerative force means "up."

The pilots are, tragically, using the fluency heuristic; they infer that they're converging on the target property—in this case, elevating the plane—because it "feels like" they are. The psychological experience feels right. We would say that the pilot made a fluency error that had disastrous consequences. But can the pilot be mistaken in experiencing a feeling of ease? No. Rather, he was mistaken in using this cue as a reliable indicator of truth: "I am indeed climbing." While the false climb illusion doesn't occur as easily or frequently as most fluency effects, it is a clear example of the sheer power of fluent falsehoods, when feeling that *p* (you are climbing in altitude) is not an accurate indicator that you are, in fact, climbing. As we will see, scientists also make use of the fluency heuristic as they develop theories to explain the world around them. And while errors in fluency don't always produce a dramatic "crash and burn," they can powerfully influence scientific explanations and theoretical development.

FLUENCY FROM PRACTICE: SCIENCE BOOT CAMP

If the G-force heuristic for aerial ascent creates a sense of *navigational* fluency, where does the fluency of a *scientific* explanation come from? The answer comes from both explanatory prototypes and the ability to recognize fluency. A significant portion of the former comes from immersion in a theory, which cuts deep grooves in our cognitive processes. For example, by the time a scientist is ready for the PhD job market, he or she has spent about 20,000 hours engaged in the professional activities of a scientist: five years in a PhD program, and at least two as a postdoctoral fellow.

All researchers are steeped in the culture of their specialization; they are specialized to the point of being insular. From X-ray crystallography to condensed matter physics, molecular pharmacology to genetics, population ecology to paleomagnetism and climate science, scientists are completely immersed in the organs of the profession, running experiments, disappearing into textbooks or research articles, engaging in professional conversation with lab members, writing grants, attending or presenting at professional conferences, and otherwise having constant contact with the norms, vocabulary, findings, and expectations of their profession. And this isn't just some egghead brand of freshman hazing. There is a lot at stake in maintaining a science that excels: millions of dollars in grant funding, new professional ties with affiliated disciplines, technological advances that improve human well-being, and, of course, experiencing the joy of discovery that comes from good theory, insight, and effort. In time, they have fully internalized the explanatory prototypes of their field, and with this completely assimilated material comes a sense of understanding through specialization.

The noted historian of science Thomas Kuhn took many opportunities to explain how the young scientist cannot escape the effects of this practical immersion into a theoretical worldview:

> At least in the mature sciences, answers (or full substitutes for answers) to questions like these are firmly embedded in educational initiation that prepares and licenses the student for professional practice. Because that education is both rigorous and rigid, these answers come to exert a deep hold on the scientific mind.[24]

Fluency is a property of every scientific (and for that matter, any highly practiced) theory that has been accepted for any significant period, whether or not it was false:

> Aristotle's *Physica*, Ptolemy's *Almagest*, Newton's *Principia* and *Opticks*, Franklin's *Electricity*, Lavoisier's *Chemistry*, and Lyell's *Geology*—these and many other works served for a time implicitly to define the legitimate problems and methods of a research field for succeeding generations of practitioners.[25]

Kuhn's description of the social institutions of science develops this connection between highly specialized, insular training and fluency. Theories, like paradigms, serve an organizing function in the scientist's mind and in the student's curriculum:

> In that role [a paradigm] functions by telling the scientist about the entities that nature does and does not contain and about the ways in which those entities behave. That information provides a map whose details are elucidated by mature scientific research.[26]

The remarkable uniformity of outlook is actually carefully taught:

> A scientific community consists, on this view, of the practitioners of a scientific specialty. To an extent unparalleled in most other fields, they have undergone similar educations and professional initiations; in the process they have absorbed the same technical literature and drawn many of the same lessons from it.[27]

Explanatory prototypes are especially prominent in highly specialized fields; the prototypes consolidate extremely technical features—often dozens of them—into a concept that is familiar to highly trained specialists who speak the same arcane language. The history of science supplies endless examples of explanatory prototypes, collections of postulated properties used to account for certain phenomena. Newtonian explanatory prototypes are a classic example: spherical corpuscles with negligible volume, capable only of elastic collisions inside a container, will interact in such a way that pressure and temperature increase when volume is reduced. We come to expect that an explanation for changes in pressure or temperature include aspects of that prototype, such as increased motion of the molecules or corpuscles, their energy, and so on. Here, the specialization-induced fluency creates a sense of understanding that repays loyalty, because it's true.

This is not always the case. Specialization in a false theory can produce hazardous fluency. For example, Aristotelian and neo-Aristotelian variants of medical theories forged explanatory stereotypes that were hydraulic, and these accounts had very little reliable evidence in their favor, and much against. Such theories might be called *fluent falsehoods.* One explanatory prototype invokes accounts for illness by appeal to an "imbalance of humors" and of conditions like excess and insufficient "humidity" in the body. There are no such conditions, of course, but there are explanatory prototypes that the theory crafted to account for ordinary experiences of symptoms like diarrhea, vomiting, and fever that can be relieved with liquids, for example. So humoral explanations, though based on a deeply false theory, made a certain amount of practical sense. But pushed too far, these explanations would create disastrous results for the poor patient. Or consider the dominant 12th-century Islamic "extramission" theory of vision, recovered from ancient

Greek accounts, which included explanatory prototypes of corpuscular light, and visual rays projected from the eyes. The medieval extramission theory enjoyed the authority of Euclid and the ancient Greeks, and its commitment to the eyes' projection of visual rays provides an easily pictured model of the rectilinear propagation of light. So once again, this theory of vision endured in part because, in the absence of a good optical theory or a biological theory of visual information processing, this view was consistent with some phenomenological theory of vision. But a fluency that produces theoretical success requires more; it depends on specialization in an accurate theory, one that explains illnesses not in terms of humors, but in terms of microbial, genetic, neural, hematological, and endocrinal sources proper to the illness. When this happens, scientists experience a sense of understanding that orients them toward success.

People are spontaneously aware of a process's fluency (or lack thereof) every time they realize that they can now listen to music while they knit, play catch while they talk, or do nothing else when they compute a difficult math problem. When we make this judgment, we are using a crude measure of how easy it is to think about something. Because we can track that ease, fluency is an adaptive heuristic of sorts that helps us to allocate precious resources to appropriate problems; it allows us to sit back and reduce our monitoring when we feel that things are going well and to take action when something feels "off." Sometimes these feelings are tied to the conviction that some statement is true or false. If the statement that the Amazon River is longer than the Nile feels right, we judge it to be true. If it doesn't feel right—if it doesn't feel like it fits with what we know or have heard—we judge it false. In fact, that feeling may even be part of that evaluation. But as we will see, the feeling of fluency is no better an indicator of (approximate) truth than the feeling of understanding.

These two phenomena place explanation at the center of a puzzle about scientific progress: scientific progress depends on accepting accurate explanations, and yet the feeling of understanding in its different flavors possesses no reliable cue of an explanation's truth. There are many possible objective properties of good explanation, and ultimately it might be true that our path to accurate explanations is guided by one, or even all, of them: unification, coherence, high probability on the event to be explained, and so forth. Ultimately, that may be true. But nature is too complicated, and the human mind too small, to keep an accurate running and weighted count. Consider a system as seemingly simple as a watershed, whose pollution could be explained by any or all of hundreds of variables, operating in different directions. Our theories don't give us direct or transparent access to the ontology that would make our explanations for the pollution true. The universe is full of such systems, so we must use shortcuts or heuristics, either ready-made or self-consciously crafted as a methodological tool, to span that distance between our sense of understanding and the ontic facts that make the explanation accurate.

Can we rely on cues other than this sense of understanding? Every serious effort to set standards in explanation attempts something more secure than this subjective cue. The Inductive Statistical approach requires that the explanation of an event confer high probability on the event to be explained. Van Fraassen's account stipulated that explainers must be probabilistically relevant to the thing that needs to be explained. The unification approach looks instead at another objective standard for appraising explanatory goodness—an explanation's ability to account for a range of different phenomena previously thought to be unrelated. The clue that the explanation is good can be found in the reduction of the number of facts we must claim are ultimate or basic. The best examples here are Newton's unification of terrestrial and celestial mechanics and Maxwell's unification of electricity and magnetism: they rely on the intuitive appeal of reducing the number of facts we must claim are ultimate. Thagard sees epistemic coherence as a key feature of explanatory goodness[28] and edges toward a more objective account of this goodness by determining how explanations maximize satisfaction of constraints like consistency, compatibility, and (positive) association. These efforts are regularly vetted in professional philosophy journals, and they suffer from their own internal problems. So it is tempting, and perhaps appropriate, to focus on factors that actually drive the acceptance of an explanation—factors like fluency or the feeling of understanding.

Consider the feeling of understanding that unification supplies. The prospect of unification creates an "oceanic feeling," a sensation of an intricate bond, of being connected with the most fundamental structures of the external world.[29] Faithful to the truth or not, this elevated feeling induces the most ecstatic pronouncements of connection with the world, of coherence and place. With some systematic study, we might be able to grasp the oceanic feeling we get when we observe a grand design that we know represents far more than we can appreciate. After all, there has been plenty written about some of the most prominent thinkers in history who have drowned in this oceanic feeling, following a bad but alluring theory to its demise. Even so, sense of understanding has been a powerful force for theoretical progress when our theory oriented us correctly.

Science has progressed because people correctly decided which explanations to accept and which to reject. And while good explanations may necessarily conform to logical requirements like noncircularity, people—scientists included—often accept explanations for distinctly psychological reasons. Einstein embraced a hidden-variable interpretation of quantum mechanics because he couldn't imagine a God that plays dice, or a universe that was irreducibly statistical. Haldane rejected reductionist accounts of heritability because he couldn't conceive that the mechanisms of heritability were physical. Copernicus tried to avoid two counterintuitive commitments, claiming of two alternatives to heliocentrism that "the mind shudders at either of these suppositions."[30] It is hard to see how this mere personal report enhances the credibility of heliocentrism. The important question is whether heliocentrism is true, not whether envisioning an alternative is too intellectually painful to bear. Yet, this is the standard so often appealed to when deciding which explanation to accept.

Suspension of belief is often called for, but we are intolerant of uncertainty, impatient about what William James called "that peculiar feeling of inward unrest known as indecision."[31] Instead, we are heavily padded with overconfidence. In fact, some of the most ill-fated enterprises were mounted and sustained by thinkers with supreme confidence in their explanations. Consider the alchemist Paracelsus, who claimed to have found the "Universal Medicine" in his *Archidoxis: Comprised in Ten Books*, on the grounds that "By means thereof I have cured the leprosy, venereal disease, dropsy, the falling sickness, colic, scab, and similar afflictions; also lupus, cancer, nolime-tangere, fistulas, and the whole race of internal diseases, more surely than one could believe."[32]

We conclude that we are zeroing in on an answer when we feel the tension of uncertainty release and when we have the "Aha!" feeling, some kind of aesthetic pleasure, the feeling that we're really working hard at learning, the feeling of agreement from a cooperative consensus, the feeling that we're acting for good/rational reasons, and so forth.[33] So the reason that fluent processes feel good may be shockingly like the reason men have nipples: it is the runoff of the evolutionary process. The feeling of fluency is mildly pleasing, associated as it is with the sense that a problem has been correctly solved because we have a surplus of processing capacity.

No one is surprised by the fact that there *are* fluency effects. Rather, they are astonished by their reach, by their ability to deter acceptance of superior explanations. For example, just as today irrelevant magnetic resonance imaging scans described in a fluent neuroscience vocabulary increase satisfaction with an explanation, false information that had an intuitive plausibility typical of a premodern age kept ancient thinkers satisfied for centuries. Aristotle explained matter-of-factly that the sex of goat and sheep offspring depends on which way the wind is blowing: "[I]f they submit to the male when north winds are blowing, they are apt to bear males; if when south winds are blowing, females."[34] Aristotle appreciates a world of many causes and asserts that the consistency of the sperm will also determine the sex: "[W]hen granular it is fertile and likely to produce male children, but when thin and unclotted it is apt to produce female offspring."[35] Twinning, on the other hand, was the result of the "richness of the pasturage" that the parents enjoyed.

The credibility of these explanations depended entirely on the fluency of authority, and the network of biological concepts that made up the constellation of ancient Greek medicine. As far as we know, there was no record made of offspring sex as a function of wind direction or sperm granularity. With similar dash, Descartes drew on the mystical mechanism of medieval physiology when he explained that birthmarks form when a pregnant woman looks at the sun, because his theory of animal energy suggested a kind of mechanical interaction between object and effect of perception: "Sometimes a picture can pass . . . through the arteries of a pregnant woman, right to some specific member of the infant which she carries in her womb, and there forms these birthmarks which cause learned men to marvel so."[36]

In this "feels right" epistemology, how does the sense of understanding feel? Sense of understanding is a phenomenological state. There is "something it is like" to have a sense of understanding. Part of this sense is one of grasping a point, of being richly located in a network of causes that normally comes from a coherent outlook. Different people have endorsed different descriptions of the sense of understanding. The most common is the "Aha!" description, like a flash of insight. The physicist Steven Weinberg defines it operationally: "Explanation in physics is what physicists have done when they say *'Aha!'* "[37] But this is too narrow an account of the sense of understanding that often comes with a good explanation. Sometimes it is a slower appreciation, or a dimmer dawning. Richard Feynman, Nobel Prize–winning physicist, likes the image of the dramatic hero, latching onto an insight with a genius so powerful that even the most complicated solution is obvious to you: "You can recognize truth by its beauty and simplicity."[38] But, of course, this is false. Feynman was not making the point that "recognize" is a success verb. Instead, he was inviting readers to suppose that this is why he kept at a problem while others gave up: because he was able to recognize truth. The problem is, beauty and simplicity are theory-dependent virtues, and not the other way around. Pronouncing a theory or statement as simple or beautiful doesn't make it so. Is a world with a God simpler than a world without? It depends, among other things, on whether you render the world more complicated by introducing another unreduced entity (God) or more simple because it can account for the behavior of more (types of) things. Einstein worried over just this trade-off: you can retain a simple Euclidean metric and complicate your physics, or have a simpler physics and treat space as non-Euclidean.

The embarrassments of this flabby pattern of confirmation are never showcased. Every medieval doctor, every ancient astronomer, and every Renaissance geologist reported with great authority the quick and unexpected path of a proud discovery. Of course, those "discoveries" killed their patients with mercury or lead, misled generations with a prolix model of orbits incapable of explaining some of the best-known observations of the time, and killed mine workers in collapses. No thorough treatment can ignore the regular occurrence of these features—conciseness, suddenness, and immediate certainty—that strike scientists daily.

And yet, sometimes that psychological experience *is* right; sometimes it is a faithful indicator of the way the world really is. In many simple cases, there is no reason to doubt the accuracy of the fluency heuristic. Fluency itself is not rare or exotic. It is the mildly positive feeling of fluency that we get when absorbed in mindful activities—from soccer to chess, from conversation to music composition. Our lives are filled with fluent hunches, feelings that whatever beliefs we happen to be entertaining are correct, or that we have the right answer in a given situation. We have endless opportunity to have these hunches, because we are expected to answer all sorts of questions all of the time. We have plenty of opportunity to be wrong, because many of these questions are about technical matters that are only responsibly answerable by arcane theoretical information, not by consulting our intuitions or impressions on the fly. Explaining the difference between a song and a poem may draw only on an intuitive, rudimentary grasp of common subjects, but not the difference between homologous and analogous structures in evolutionary biology, or the learning cycle of the caudate nucleus that produces the feeling of fluent hunches. People have reliable hunches about their enjoyment of eggplant fried rather than ratatouille, or whether a local deli is farther by one set of roads than another. Appliance technicians have reliable hunches about whether a particular kitchen product malfunction is mechanical or electrical, and plumbers have reliable hunches about whether a slow drain is the result of a turn in the pipe that is too tight or a pitch that is too slight. But even then, there is no reason to rely on a hunch when you have a camera that you can snake into the pipe.

Scientists may have reliable hunches about whether an experimental task will control for order effects, whether a complicated chemical reaction is exothermic, or whether a galaxy's rotation is relatively fast. So the hunch by itself is not the enemy of sound inference, but rather the hunch is oriented fruitlessly, and perhaps needlessly, by a poor theory.

PSYCHOLOGICAL FLUENCY: NATURAL KINDS, PROTOTYPICALITY, AND BEAUTY IN THE MEAN

Science is about special kinds of classes in nature. It is not interested in phony hybrid objects like a salamander 54 miles southeast of the Liberty Bell or elements with atomic numbers that are the sum of pets you have. Instead, science is interested in "carving nature at its joints," to use the apt Platonic phrase further popularized by Quine. Once the theory carves those joints, it has exposed Natural Kinds, objects in nature that play a taxonomic role in a mature, working science. What makes these kinds natural is that it is nature itself, not human practices of categorization, that fixes whether an object belongs in one class or another. Species and crystals, copper and maple trees, and visual transduction and lexical priming are natural kinds. Natural kinds are the very stuff of science, and yet many of them are not the compositionally inert, timeless objects science has promoted as its image. These natural kinds are no more timeless than an evolutionary lineage, no more frozen in composition than the deviation of isotopes and the variability of neurally plastic processes.

As a psychological description, we recognize natural kinds by looking for their central tendencies, the most representative member of a category or the average of all the members belonging to a category. Psychologists of concepts call this the prototype.[39] The preference for prototypes is robust: people display it for living categories like human faces,[40] as well as fish, dogs, and birds.[41] But the drive is so strong to find the center that humans do it even with categories of nonliving objects, such as color patches,[42] and even artifactual objects, such as furniture,[43] wristwatches, and automobiles.[44] So robust is this drive, in fact, that objects that clearly do not admit of gradations receive this treatment from humans. Whole numbers are one such example: certain odd numbers are reported by people to be "more odd" than others.[45] Who would have guessed that 7 and 13 are the oddest of odd numbers, and 15 and 23 the least odd of them? Or that 8 and 22 are the evenest even numbers, and 30 and 18 the least even?

This drive leads to all sorts of different prototypes: in the United States, robins and sparrows are prototypical birds, followed by birds of prey, then poultry, and finally, the clumsy flightless ones.[46] There are prototypical molecules (taken as normal compared to their isotopes), as well as prototypes of a biological species, of a disease, or of a mountain range. Years of training make these prototypes present-to-mind, what psychologists call "chronically accessible"—that is, what becomes "prototypical" is that which is easiest for the brain to process. And in science, just as in everyday life, explanatory prototypes that free up processing space in the brain are deemed more attractive, and more accurate, whether or not they actually are. Fluency feels good, and disfluency feels bad.

HOW CAN FLUENCY'S SENSE OF UNDERSTANDING CAUSE POOR EXPLANATIONS?

Brain imaging evidence is now a potent part of explanatory prototypes in cognitive neuroscience. A novel series of experiments by Deena Skolnick Weisberg and her colleagues show that nonexpert consumers of behavioral explanations assign greater standing to explanations that contain neuroscientific details, even if these details provide no additional explanatory value. This "placebic" information produces a potentially misleading sense of intellectual fluency and, consequently, an unreliable sense of understanding.[47] This extraneous information, especially neuroscientific information, gives people a mistaken feeling that they have received a good explanation. But why would placebic information (rather than, say, false or shocking information) create a sense of fluency? This question goes beyond the scope of Weisberg et al., but the conceptual connections are easy to trace. Placebic information has characteristics that promote the feeling of intellectual fluency. The technical vocabulary and causal taxonomy of placebic neuroscientific information might activate conceptual representations contained in "true" psychological and neuroscientific explanations. Thus, irrelevant information can still convey the good feeling of fluency experienced when we assemble and process an explanation.[48]

In their second study, Weisberg et al. created bad explanations: ones that were circular in nature, thus violating a fundamental formal and logical constraint of explanation. Because the bad explanations were just circular restatements of the interesting effects, they were not explanatory. For example, one circular "explanation" for an error in perspective taking states that the error "happens because subjects make more mistakes when they have to judge the knowledge of others. People are much better at judging what they themselves know."[49] This "explanation" is just a restatement of the initial clause, and it assumes the truth of the very thing it purports to explain—that people are relatively worse at judging what others know. But subjects found the circular explanations more "satisfying" when they contained neuroscientific vocabulary, or "neurobabble."

These two experiments cross quality of explanation (good vs. bad) with presence of neuroscientific information (with vs. without). Overall, nonexperts found good explanations significantly more satisfying than bad explanations, and explanations with neuroscientific information more satisfying than those without. In addition, bad explanations with neuroscientific detail enjoyed a neurophilic premium; neuroscientific information produced a special boost in perceived accuracy. Neuroscientific vocabulary delivers additional fluency over behavioral explanations, whether actually useful or not.

Most people would like to think that we are not so gullible about bad (circular) explanations with pretty neuroscientific words—that a little training in cognitive neuroscience would make us critical enough to defeat such seduction. But training in that field does not help much—similar to novices, students enrolled in an introductory neuroscience class appraised explanations with neuroscientific information as more plausible than those without. Furthermore, bad explanations benefited more from placebic neuroscientific information than did good explanations. Experts were less sensitive overall: they didn't recognize how bad the non-neuroscientific explanations were, and they were less impressed with the good neuroscientific explanations. There are similar effects on experts in other domains, in which experts moderate their assessments. Experts might be less hoodwinked by neuroscientific explanations for the same reason that improving a person's skills reduces their overconfidence about them.[50] Except for truly expert subjects, placebic neuroscientific-explanation-primed conceptual vehicles deliver feelings of intellectual satisfaction. And placebic information is not the sole supplier of fluency; this sense of understanding is conveyed by several well-documented psychological effects, such as the feeling of knowing, illusion of explanatory depth, and tip-of-the tongue experiences.[51]

Novices accepted circular explanations because the reductive neuroscientific details sounded so credible. And scientists routinely accepted bad explanations when their arcane vocabulary was used. The lesson of this research, then, is general: explanations, bad and good, are routinely accepted for non-truth-related reasons. This is all the more reason to be cautious about our gut reliance on our sense of understanding.

METACOGNITIVE CUES OF MENTAL EFFORT

Scientists theorize a lot. They build models of chemicals, of population change, of the fluidity of materials, and of immune response. They detail plausible causes for effects that are not yet well understood. Did the population of garter snakes decrease as a result of less prey, of toxins affecting fertility, or of increased predation on babies? Is there more than one of these causes, or still others that don't appear on this list? And scientists, of course, test hypotheses. They might, for example, identify a region in which the slug population is treated with the poison metaldehyde and hypothesize that those regions will have a lower garter snake population, or that garter snake autopsies will reveal high concentrations of metaldehyde. This pattern of reasoning flows easily from prominent concepts in herpetology, theoretical/causal models of fertility in reptiles, the mechanisms that might be affected by different forms of toxicity, and so on. For the specialist, these paths of thought produce constant low-grade pleasure. Treading familiar and engaging territory feels good, as though the trek will reward thorough examination.

Theorizing, hypothesizing, testing, and categorizing—these are the activities that scientists perform, the things they do. And they do them constantly. In their unending efforts to categorize or hypothesize, as in ours, fluency leaves them with the feeling of rightness. But sometimes our paths of thought are laden with obstacles, and nothing feels easy. When retrieval feels fluent, our metacognitive monitoring takes this experience and delivers its assessment of reliability. We use our metacognitive powers to take our internal temperature. How much attention are we paying to the problem we are trying to solve? Are we focused? Do we have the skills to solve it? How easy are the equations for me to solve? These processes reflect upon thinking. Yet it is an open question whether our thermometer is accurate when we ask, "Am I rushed, distracted, frustrated, or competent at this task?"

It may seem that fluency can be used to account for too much about explanation—the pursuit of understanding and the reward of theoretical success, an effect of professional training and a source of organization. But fluency doesn't guarantee success; our judgments of fluency are more discriminating than that. True, we evaluate feelings of fluency and seek these feelings, but we are also averse to feelings of *dis*fluency. Shah and Oppenheimer found that people were more likely to choose fictional Turkish stocks with names that are easy rather than difficult to pronounce,[52] while Novemsky et al. documented that consumers defer a purchase, or stick with a familiar option, when the item has a disfluent name. Names like Artan, Dermod, Tatra, Pera, Ferka, Kado, Alet, and Boya were fluent. Names like Lasiea, Siirt, Aklale, Taahhut, Emniyet, Dizayn, Luici, and Sampiy were not. Once again, metacognitive judgments are controlled by the unreliable cue of the fluency with which we access information.[53]

We have limited cognitive resources, and the fluency heuristic allows us to appropriate these resources at the proper times—times where the task is difficult and demands deliberative concentration. But fluency errors happen all the time, no matter how convinced we are of our correctness. We can be misled by a "feeling of correctness" whether we are at the controls of an airplane or an explanation.

FLUENCY AND EXPLANATION

Our fluency heuristic trades, and sometimes equates, perceived simplicity for accuracy: it just *is* the substitution of the deliberate and analytic determination of truth with the feeling of ease. But while the case of the False Climb, in which the feeling that the pilot was climbing was used as evidence for the truth of the claim "I am indeed climbing," culminates with a fiery crash and burn, examples from the history of science more subtly illustrate the dangers of substituting truth for ease of processing.

This feeling of fluency can be caused by the most primitive repetition, as well as the highest flights of sustained theoretical learning. Historian of science Thomas Kuhn recognized that members of a scientific community "have undergone similar educations and professional initiations; in the process they have absorbed the same technical literature and drawn many of the same lessons from it."[54] All that time and effort has a psychological impact. Humans don't have that kind of intensive and prolonged exposure to anything without developing powerful habits. But these habits are still acquired with a folk or naïve "feels right" or "makes sense" epistemology. This fact of cognitive entrenchment produces a grand puzzle: if a feeling of understanding accompanied both good and dreadful theories in the history of science, how did we ever move to a better theory, rejecting the hydraulic view of stroke and eventually embracing a neural one, or discarding an impetus theory of motion for an inertial one? The answer is a combination of hard work and a good measure of luck or contingency. These features are complementary, and there is no recipe for the proper balance of hard work and luck. But that is as it should be; the initial quality of theories, scales of complexity, and distance from truth all vary so greatly from theory to theory. The grain of truth in Thomas Edison's adage that genius is 1 percent inspiration and 99 percent perspiration is just this: it is difficult to advance a field without engaging vigorously with its subject matter. But vigorous engagement is not a sufficient condition for advancement: we have already seen that active commitment to false theories like extramission can stymie a field for centuries. Once you have a sufficiently good theory, however, vigorous engagement is the only route to theoretical progress. So you have to study the field, hunt down and organize the data, formulate hypotheses to test, and design tests that are not prematurely specific or pallidly general, while controlling for the right, plausibly interfering, artifacts. These practices help, and are helped by, hard, focused thought about the causes named in the hypotheses and about the norms in affiliated fields, as well as by spirited conversation with people thinking at least as hard and disposed to share their thoughts. With the proviso that you have embraced a theory that is good enough, it will be a reliable strategy to infer the truth of predictions it makes and explanations it offers; that is, the theory, in general and in the long run, is more likely than not to produce success and progress. So while fluency in a false theory is a terrible drag on progress, fluency becomes a virtue when your theory becomes good enough.

FLUENCY EFFECTS GROUNDED IN THEORY AND OBSERVATION

In addition to explaining the success of good theories, scientific realists should be able to explain the practical success of false or superficial theories. Let's return to the example of Aristotle's theories of reproduction. Our modern view of human reproduction treats sex as genetically determined. It makes no room for sex determination by wind direction or sperm consistency.[55] In fact, the modern view can explain *why* Aristotle's factors are irrelevant. The fluent, modern account uses terms like "chromosomes" and "genes," "ovum," and "division." But these uses are the result of *theoretical* achievements, having gotten things right about factors that are *unobservable*. Yet there were plenty of practical successes in the realm of animal husbandry or breeding before these modern genetic accounts. These successes, however, were not theoretical—they were not the result of getting things right about unobservable mechanisms or processes. Instead, they were achievements in the tracking of correlations among *observable* properties.

For example, consider what knowledge is available about the offspring of dogs, just from casual observation of their properties. Stray dogs, and their mongrel puppies, have been commonplace ever since humans domesticated dogs. Humans bred their dogs for specific phenotypic traits—most often to hunt or fight—and those traits could be easily observed. Tracking those observed properties, humans no doubt noticed that mongrels are less likely to have short muzzles than stray dogs. Why is this?

Well, a long, pointy muzzle is a dominant trait, and a short muzzle is recessive (see Figure 2.2). When one mates with the other, the offspring will have a long muzzle. To have short-muzzled offspring, both parents have to be short-muzzled (or carry the brachycephalic gene).[56] Because breeds with short muzzles are less common, and the breeds themselves are not large in number, it would be surprising indeed for a stray Pekinese to just happen to have bred with a Boston Terrier, a Boxer with an English Bulldog, or a Bull Mastiff with a Rottweiler.

It may be that European communities in 600 AD enjoyed a sense of understanding about the persistence of the long-muzzle trait in an uncontrolled stray population of dogs. They might have recognized that the offspring of a long and short muzzle had a long muzzle. But in fact, they had no idea what the underlying causal mechanisms were, no idea why the short muzzle is so rare in a population of mutts, and no idea of the factors that might suppress the expression of a trait or what mechanisms carry the muzzle information.

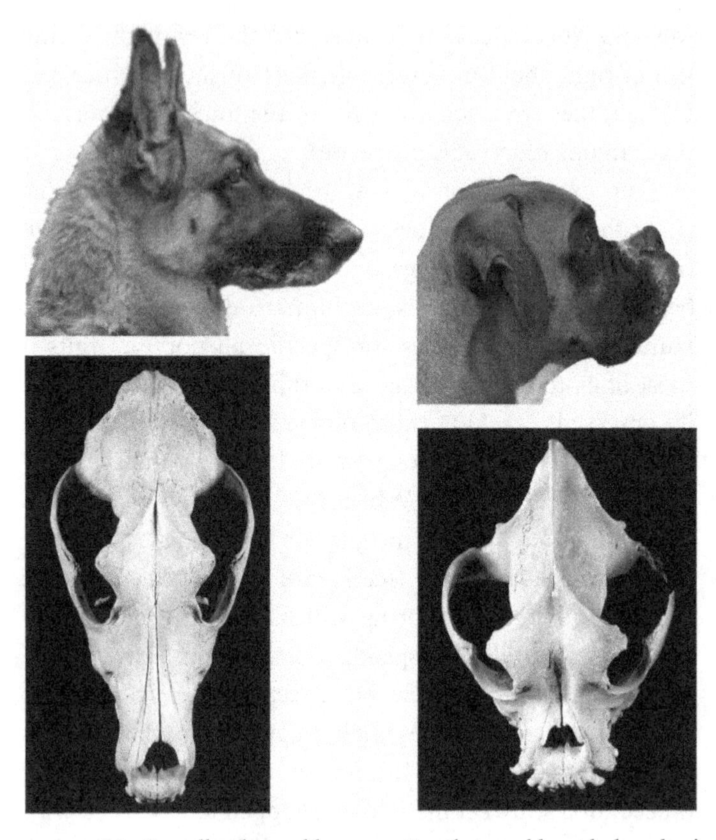

FIGURE 2.2 Casually observable properties that could guide breeders' pursuit of selected phenotypic traits. Courtesy of Mary Bloom.

They began with a superficial understanding of some observable aspects of heritability, in the form of knowledge about correlated properties. And they could use this understanding in practical ways. But their poor theory of unobservable causes would traipse across the causes that actually account for the absence of brachycephaly in mutts.

Fluency effects, then, are not necessarily grounded in knowledge of theoretical entities; the earliest examples of controlled dog breeding show us that we can achieve practical successes in the absence of theoretical development. No one doubts that *something* unobservable was being manipulated in the case of dog breeding, but the practical craft of dog breeding did not rest on knowing the characteristics of those unobservables.

It is a real mark of progress to even know that a fluency effect is at the bottom of the acceptance of a false explanation or, as in the case above, what the true causal mechanisms underneath a superficial explanation are. But diagnosing the error in false or superficial explanations isn't going far enough; it is important we correct it as well. How should we go about fixing this cognitive issue? Perhaps we can pry apart fluency and acceptance by extracting lessons from other cognitive heuristics prone to failing. Debunking the overconfidence bias is one such example. When this bias is activated, we infer the probability of our correctness from the strength of our feeling of confidence about a factual matter. As it turns out, we are routinely overconfident. To undo this effect, psychologists have given the overconfident subject a "consider the opposite" task, in which people are asked to generate specific hypotheses about how we might be mistaken. This laborious exercise, too intrusive to implement as often as it is warranted, has been shown to reduce but not eliminate the effect.

Or we might try to activate a deliberate psychological process or suppress an intuitive one.[57] Small-scale, long term follow-up studies have tried to test the sustainability of these two simple tricks. These studies would ask whether people continue to perform better than controls after spending hours holding a pencil in their teeth to induce happiness from smiling,[58] or ask whether people continue to solve problems better than controls after months of furrowing their brows while problem solving. Yet even if such internally generated improvements proved sustainable, it seems like they would be onerous and intrusive. While there are a number of "cheap tricks" that might improve the accuracy of our reasoning strategies, it seems almost too fanciful to believe that instructing people to furrow their brow while they work could improve their syllogistic reasoning abilities. But more important, we have never examined their long-term fate. It would be desirable to create procedural strategies that replaced such tricks with other reliable processes less prone to fatigue and inattention, that is, processes demanding less concentration and motivation to apply. On the other hand, having an accurate explanation—being right—provides exactly the sort of reliable procedure we need.

When you don't know whether a theory is accurate, of course, you have no special reason to interrupt fluencies. In that case, the natural environment of scientific theorizing, rather than deliberate efforts at interruption, may intervene to point out that a fluent theory is false, or deeply problematic. In the same way, an accurate explanation may come out of the blue; it may appear with little historical precedent distinguished from explanations that were nonstarters. So, while cheap tricks that activate deliberate processes may clear the palate, preparing the way for less biased reasoning, people are unable to correct these biases spontaneously. If we have any say in their reduction, it must be achieved in the conditions we arrange for scientific experimentation and theoretical reflection. Peer review, for example, does not guarantee impartiality, but it blocks systematic tendencies to advance and sustain a poorly supported view. Critical scrutiny from peers may sting, but it will slow you down long enough for you to soberly consider that a familiar view may be false. Or it may happen naturally. Scientists fluent in the use of their paradigm's or theory's concepts and terminology will generally continue to understand the world in those terms unless the world steps in and corrects them abruptly, as it did when Boyle faced corpuscular mechanics, Semmelweis looked at the lopsided deaths that varied by handwashing procedure, and Jenner ventured the basis for the cowmaids' freedom from smallpox. Performance improves dramatically when the world steps in, and experiments are one way to issue that invitation. And sometimes without experimentation an event is so robust that it is impossible to sustain an illusion or it is absolutely clear that an existing belief has been hidden or protected from decisive feedback. Contingency, then, can be a powerful corrective force for scientific progress, in addition to the help that experimentation can offer.

CONCLUSION

The origin of explanatory fluency orients us toward two conclusions. The first is that despite the occasional, purely instrumental or empirical descriptions of the world (like the observational account of dog breeding) that bring practical value, most of the history of science was a false climb. The second is that, in the life of the working scientist, mastery of the dominant theory is like fluency boot camp. A core consensus about protons, black holes, genetic drift, and ocean acidification reaches thousands of scientists through textbooks and conferences, and gets reinforced in lab meetings and referee reports that punish deviations from the fluent canon. And fluency, we have seen, feels good: psychological processes of learning do not stand in complete isolation from all others. After all, learning is motivated by experiences like pleasure and confidence and is guided by sensitivity to feedback. The sense of understanding is enjoyable, or positively hedonically marked. It didn't *have* to be; it is just a contingent fact about humans that pleasure is the powerful motivator here. There might have been other routes to fluency—efficiency perhaps. But efficiency is not, by itself, motivating. The promise of good feeling is.[59]

The fact that fluency is marked by a *feeling* may come as no surprise. For psychological beings like ourselves, a heuristic would have to involve thoughts and accompanying feelings of *some* kind. After all, there is likely to be *some* kind of internal, phenomenal state that goes along with psychological processes that have outcomes perceived or coded as good or bad. My view does not recoil from the idea that there is a feeling associated with the heuristic, or even that the feeling might be the sole basis of the heuristic, but rather that the heuristic is not a reliable mark of truth. The goal is not to have a world without heuristics, or sense of understanding, but one that replaces unreliable dependence on them with guidance by sound cues.

This story of psychological fluency accounts for the acceptance of many intellectual views, not just scientific explanations. But here we have focused on scientific explanation. Training in a theory cognitively shapes the scientist's explanatory prototypes. And so, explanations are the products of an age: Ancient Greek and Islamic humoral medicine sought and favored explanations with concepts like liquid and heat; those explanations were more fluent than those of either the less practiced theories or the undiscovered ones. Ancient Chinese explanations for a comet's appearance invoked an imbalance of yin and yang, Ancient Greek ones the dryness and fatness drawn up from the earth. And all the while, the available explanations felt good. Candidate explanations had a leg up if they conformed to a familiar explanatory pattern, using concepts that formed a kind of prototype of an explanation. The Newtonian explanatory prototype, found in everything from pressure to planetary orbits, tended to appeal to ballistic particles subject to a handful of mathematically tractable forces.

Psychological fluency is a conservative process. Scientists experience a feeling of understanding when crafting or listening to explanations that make sense of an existing puzzle by connecting familiar concepts. This experience originates from stable and chronic associations between the stable environment and heavily practiced theoretical activities of scientists. Any theory of scientific progress tied to so conservative a process has to account for the occasional acceptance of *novel* explanatory prototypes that come with the acceptance of new theories, given that the history of science is punctuated by grand and sudden changes, revolutions of outlook. There is a credible account to offer of how that occurs, but we save that for later in this book.

If your flight instruments are accurate and reliable, you will actually be descending when the instruments say so. You may still *feel* like you are climbing, and knowing that you aren't doesn't make the feeling of ascent go away. The same goes for the history of science. Premodern scientists had the sense of understanding, the feeling that they were tracking the truth, but in the grip of theories of miasma and epicycles, of humors and mystical reproduction, their explanations conveyed a sense of understanding that was counterfeit. They were at best moving sideways, and at worst, heading downward, into the darkness. Plain and simple, when it comes to arriving at a good explanation, there is no substitute for having a good theory. And, given how unreliable fluency is as an indicator of truth, particularly in complex theoretical matters, the best explanation for theoretical progress in modern science is accuracy rather than fluency. In the dark, navigating by our feelings, our sense of understanding, is the way of disaster. No matter how they make us feel, accurate scientific theories point us along an upward trajectory. We need only learn to fly by those instruments and avoid the false climb.

3

THE BIOLOGY BEHIND THE FEELING

The Neuroscience of Explanation

INTRODUCTION

When medieval doctors tried to adjust the patient's humors, they expected the patient to improve. Aristotle expected the direction of the wind and the coarseness of sperm to determine the sex of offspring. Soothsayers and astrologers expected to be able to anticipate future events, famines and wars, marriages and deaths. These expectations rely on theories to explain why such methods were being used. Despite all expectations, these efforts failed. When your theory is bad enough, it will fail most of the time. So how do people manage to persist in believing a bad theory in the face of failed efforts?

The main reason is that true theories and false theories look the same; people can't tell a true theory from a false one just by looking or by casually examining their properties. Instead, we use heuristics, mental shortcuts, to generate an overall impression of a theory. Is the outlook or theory plausible given the other things I believe? Does it successfully predict some events I want to know about? Does it help me to explain some events, or solve problems, that I couldn't before? In short, did it lend insight or help me to learn? This chapter provides some of the psychological backstory for the account of explanatory fluency already presented, and the similarly cyclical structure of both learning and the motivation to explain, that we will find in Chapter 4 on inference to the best explanation. The explanatory strategies of scientists are informal methods, to be sure. But those are the methods that got us here. So we should learn a little more about them.

INSIGHT: FEELING OR SUCCESS TERM?

The basis of all learning is a cycle of search and reward: the satisfying feeling of understanding is a reward that is motivating. Some human learning happens without such direct motivation, but then again, humans are known for their ability to delay gratification. But most of the time, there is at least the *promise* of gratification. Some philosophers claim that the desire for understanding, for knowledge, and for moral goodness moves us to action, to motivate our search for truth. Our reward? Peaceful reflections upon our own conduct. Explanatory understanding may be a lofty goal. But that illuminating feeling presses us to *track the truth*, as that is perhaps the most reliable way to get that good feeling again.

However, that illuminating feeling can also occur when we believe what is false. If only there were a reliable cue when you believed the truth, a mark of accurate belief. Perhaps the answer lies in the ability to discern, in having insight into what is true? Two insight researchers, Mark Jung-Beeman and John Kounios, describe "insight" as a "sudden comprehension—colloquially called the 'Aha! Moment'—that can result in a new interpretation of a situation and that can point to the solution to a problem."[1]

Here we can see how those studying the phenomenon of insight in the empirical sciences directly conflate the notions of truth and phenomenology. The term "insight" is what philosophers call a "success word" (like "hit" or "see"). Ask a few dozen people whether the term "insight" implies correctness. You will find that people believe that it does. We reserve it for cases in which people make a *correct* judgment. Consequently, we don't say that astrologers had the insight that Jupiter's position at your birth confers wisdom.

When talking about insight and Aha! moments in this manner, it can sound as though researchers believe they have identified a kind of brainstate that precedes the *correct* solution to a problem. This is not science fiction. Well-known insight researchers report using electroencephalography (EEG) and functional magnetic resonance imaging (fMRI) "to study the neural correlates of the 'Aha! moment' and its antecedents. . . . Insight is the culmination of a series of brain states and processes operating at different time scales." Induce that brainstate and you are more likely to get an Aha! moment that leads to a solution: "Elucidation of these precursors suggests interventional opportunities for the facilitation of insight."[2] But there is an easy way to avoid the misleading impression that interventions can facilitate correctness: when all you mean is that a certain subjective experience accompanies a solution, use the expression *feeling of insight.* I can *feel like* I have insight into, and understand, how a radio works or the causes of the War of 1812. But I can be wrong.

The prepared mind is in a brainstate that, when present, is correlated with a subjective report of a *feeling* of insight. In some of those cases, those that possess that brainstate correctly solve the problem. In other cases, they have the feeling of insight together with the fact of failure. Consider the Remote Associates Task,[3] which is a test used to gauge creative problem solving. The research subject is presented with three words such as "baby," "spring," and "cap" and is asked to come up with a one-word solution that can combine with each of the three words to form compound words or compound phrases. Here, the target word or solution is "shower." Only when the research subjects generated this solution were they asked the self-report question: did you experience insight when solving the problem?

But here we come across a blind spot: there may be cases where the subjects have that same "feeling" that the answer came to them suddenly but they generated the incorrect solution. For example, I may believe that the target word for the task in the previous example is "bottle." This word can be combined with two out of the three words and is therefore likely to be reported as a solution by research subjects ("baby bottle" and "bottle cap"). But insight researchers are not interested in these cases of "faux insight." It is interesting to note that we can only discuss cases of faux insight if the notions of phenomenology and accuracy are already conflated—as they often are by cognitive neuroscientists. If "insight" refers to a feeling, it makes no sense to talk about cases of faux insight; we cannot be wrong about *whether* we have a feeling, only about what the feeling *indicates*. As it stands, the research shows something about the feeling of insight, not the state of accurate judgment or true belief. We must take caution and remember that feelings of insight do not necessarily track the truth. The research doesn't demonstrate that a unique area of the brain is associated with *solving* the problem, but rather with *self-reported insight or lack thereof.*[4] When a specific area of the brain lights up, people are more likely to say of a solved problem that they achieved it with insight rather than analytically. Having done so, however, does not show that you are any more likely to be correct when solved problems receive Aha! self-reports.

There is no reason to doubt that a self-reported Aha! experience is associated with a particular type of brainstate. And it is crucial to know that these self-reports of insight can often be correlated with false belief as well. Recall, in Chapter 2 our pilot was in the grip of a false climb. His brainstate told him he was headed up, up, up. But he was headed down. And the phenomenology of this experience was supercharged; the greater his acceleration toward the ground, the more it felt like he was headed up, struggling against gravity. Because reliable accuracy depends on causal relations between states within the skin and those beyond it, the structure of the brainstate can't be sufficient for accurate belief. In the case of the *false climb*, the pilot's brainstate—a representation of how his head relates to the world—is disastrously inaccurate.

The visceral nature of the brain's activity is palpable. Its behavior seems autonomous, at times uncontrollable by us. Virginia Woolf put it this way: "My own brain is to me the most unaccountable of machinery—always buzzing, humming, soaring roaring diving, and then buried in mud. And why? What's this passion for?"[5] Why are we driven to know, to understand? What is the function of the mind's teeming activity?

Woolf's question deserves an answer. Every psychological experience has a biological story behind it. The story I will tell in this chapter shows that there is only the remotest relation between the good feeling we get from an explanation when it fits into our system of beliefs, and the facts that make an explanation true. This tenuous relation is usually characterized as causal: it is a relation between the conditions in the world that normally cause the phenomenological experience in question and the conditions in the world that make the representation accurate. In this telling, the truth conditions for the belief must produce the feeling, and the feeling must carry a distinctive hedonic tone, such that it can be systematically distinguished from other feelings. No other condition can produce that same quality of hedonic tone, say, a condition that would *not* make the belief true.

But in reality, for the feeling of understanding to be a reliable indicator of accurate belief, a host of processes and items inside and outside the brain have to line up. Our biology is the basis of all of these psychological processes. We should not be surprised by the biological embodiment of our psychology. But we should also be unsurprised that our experiences are explained by what is *around* the brain, the kind of world we find ourselves in. Insight and tip-of-the-tongue experiences happen in the brain, but they are often explained by what is *around* the brain, such as contextual or environmental cues. The same goes for the feeling and fluency of understanding.

Despite the simplicity of the neuroscientific story of "insight states," this delicate network is hard to assemble: at least four things have to align just right. First, you have to be in an environment that's congenial in all the right ways. Let's take an example. For the *feeling* that you understand the causes of infectious disease to faithfully indicate that your beliefs are correct (and you do indeed understand), you must be in cognitive possession of the *correct theory* of infectious disease. Infectious disease must be transmitted microbially (rather than, say, in a "foul cloud," as medievals thought). The microbes must be able to survive in saliva, blood, breath, and so forth well enough to be transmitted in that medium. They must reproduce at an appropriate rate. They must be sufficiently resistant to immune response to survive.

Second, we are aware that topics we understand are easier to think about than ones we don't. This awareness is a metacognitive effect; we track this ease when we reflect on our own thought processes. The metacognitive effects enjoyed from processing fluency have to be carefully balanced against the compulsive need for closure, the feeling that you can stop explaining; fluency gives you flow, but you have to know when to end the explanation. The sense of understanding conveyed by processing fluency may cause you to end inquiry too soon. But the seemingly boundless number of causal mechanisms in, say, the chemistry of infectious disease related to both microbe and host can prompt explanations that go on explaining forever.

Third, for truth and sense of understanding to align routinely, you have to have enough workspace—enough "room in your head," together with well-enough articulated models and measuring instruments—to compute and double-check the magnitudes and directions of the many causes you are entertaining.

Finally, the biology of the reward cycle must be sustained by the assemblage of mechanisms connecting the satisfaction, the sense of elevation, you get when you feel an absorbing belief is correct, to the dopamine reaction of anticipated reward. This doesn't just happen; it has to be carefully taught.

The pilot's feeling of upward trajectory tracks the truth when their otoliths are being pushed toward the back of the head because the plane is *oriented upward*, and wanders aimlessly around the truth when that push comes from downward acceleration toward the ground. When the causes don't line up with our measurements, the sense of understanding is a false climb, decidedly not a reliable cue to the truth.

The intricacy of these conditions holds for nearly any theory of routine complexity. Take the example of how the feeling of understanding tracks the truth when dealing with the kinetic theory of gases. The particulate nature of molecular motion "makes sense" of the fact that diffusion takes longer per unit of distance in open spaces. Understanding the diffusion of sulfur dioxide across an open room must depend on the fact that you understand why the velocities of molecules in the gas are lower than those in a closed container. Soon enough, we can simply consult our sense of understanding on the matter.

Why do we place such stock in the sense of understanding? Our finite minds turn to simple rules to process the myriad things we hope to understand, rules that work in some environments and not others. This nest of causes imposes a hefty distance between the Aha! and real understanding. It would be arrogant to suppose that our homey vocabulary is rich enough to characterize the direction, weight, and character of these causes, just as it would be to suppose the adequacy of our folksy vocabulary to describe all of the items that mediate quantum phenomena and macroscopic effects.

Humans sometimes have that sense of understanding that correctly tracks the real world, but, if this chapter is right, not nearly often enough. True, there is a lot of space between perfect and poor correlation. But that is no reason to close our eyes and go with our gut. We may not be able to say exactly when the correlation between the sense of understanding and the real world is strong enough to warrant trust. But we have enough knowledge of the voluminous and systematic ways that people err in trusting this correlation that we shouldn't be guided by it generally. We are smart enough to appreciate the challenge of explaining really big problems, but at the same time we make remarkably simplistic and mechanical appeals to the mental processes and worldly states that make understanding possible, as though only a handful of mechanisms are involved. And that's not surprising. In so many important matters, we are absolutely in the dark about just how complex each process and state is that eventually leads to understanding.

THE PLEASURE PATH

There is a scientific, biological explanation for the desirous drive behind explanation. This desire stands at the center of a complex switching station in the brain. But the impulse to explain is more than a simple desire. It is fed by different tributaries. It can spring from natural curiosity about the topic, a personal desire to resolve uncertainty, or an appreciation of the aesthetic pleasures of explanation. But desire is still there, driving our need for explanation.

What does it mean to say that we have a desire for something? After all, there may be as many kinds of desires as there are people. Let's remember how many different kinds of desires there can be: We can desire to eat, drink, and eliminate bodily wastes. We can crave tobacco, shopping, a game of blackjack, or chocolate. We may feel the urge to play basketball or chess, to read a novel or sing. We may have desires that can only be satisfied months or even years from now. We may want to visit Sicily next winter, or set foot on every continent. Our desires may be remote in conception and in time. We may have a desire that a friend's job interview go well, a desire to get a job to avoid bankruptcy, or the remotest and overarching of desires, to have a life well lived.

Philosophers are in the business of clarifying familiar concepts like belief and desire, of working out the consequences of different visions of each. But as so often happens, there are too many theories of desire to hang a single explanation on, and they are too insensitive to the neuroscience that underlies the complicated subjective experience of desire. Instead, they are at home with Hobbes's simple account of desire. With it goes an unadorned, desire-satisfaction view of well-being: "[W]hatsoever is the object of any man's appetite or desire, that is it which he for his part calleth *good*: and the object of his hate and aversion, *evil*; and of his contempt, *vile* and *inconsiderable*."[6] That is, if you like it, it's good; if you deplore it, it's bad.

But that can't be the whole story. We don't usually conclude that just because we like something, it is good. What most people do is begin with a version of Hobbes's story of desire and fancy it up by talking about interests or informed desires. And while we know it isn't as simple as Hobbes makes it sound, it is hard to find a much more nuanced account.

Philosophers, poets, and fiction authors write eloquently of the nature of desire. Regardless of the fact that they are not of one mind about the nature of desire, all of them are in the dark about its mechanisms and structure. There are highly intellectualized versions of desire, and others tied to emotions associated with reward and punishment.

These subtle thinkers tussle with the phenomenology of desire—how desire feels, how its satisfaction is experienced, and so forth—but phenomenology is a smoky window on biological processing. The neural activity of desire is subcortical and so is not introspectible. It lies beneath our immediate awareness. The two processes of wanting and liking are the most important expressions of our subjectively experienced desires and pleasures.

WHY LEARN? BECAUSE LEARNING FEELS GOOD

The simple sea slug, *Aplysia*, cannot explain, but it can learn. It can learn, through classical conditioning, how to retract from a noxious stimulus and, through operant conditioning, how to bite.[7] If an *Aplysia* can learn, then the basic requirements for learning can't be too great. To learn, an organism must be able to anticipate. Then, the organism must be able to represent the contingency between that action and the ultimate outcome. In humans, this anticipation may ascend to the status of an expectation or a primitive prediction that a behavior will lead to a reward or punishment. The feeling is good, and we want more of it.

Dopamine plays an important role in the learning process—that much is a settled issue, though the fine structure of its role is more controversial. It is released by rewarding experiences. But it is also released when we *expect* reward. The caudate nucleus, located in the brain's basal ganglia, is like the head teacher. It allows systems downstream to clearly distinguish positive and negative feedback by increasing its activity. The caudate is in the middle of the brain, the ancient part of the brain, and also in the middle of the learning and memory action. Often characterized as a "reward center," the caudate nucleus feeds information to the higher function of feedback processing, so that someone who received praise for a correct answer recalls the rewarded features, and one punished for laziness or deception does not repeat the action. It is a crucial contributor to learning because it uses the salience-creating effects of dopamine to tag streams of input as positive or negative. This is an invaluable service to the metacognitive system, our decision-making system, which has to sort feedback into "positive" and "negative" so that our memory system can ultimately deposit this event in a form that can be usefully retrieved. Suppose you face punishment or expect monetary rewards for learning a task, so you really want to get things right. Higher motivation will turn ordinary learning conditions into a setting supercharged with instructive cues.

WE WANT TO UNDERSTAND AND WE LIKE TO UNDERSTAND

There are two neurological systems that underlie motivation: dopaminergic and opioid. As the terms suggest, the first transports dopamine and the second, opiates. The focuses of *wanting*, in the mesolimbic area of the brain, and liking, in the nucleus accumbens, are constantly engaged in a complex circuit of neural crosstalk with functions of the higher cortex, like attention and memory. The most reinforcing behaviors produce great pleasure and gratification. In addition, dopamine neurons have a special property: they can predict future reward. This ability makes them invaluable in learning. They send signals to the basal ganglia's striatum, and this changes the pathways themselves. These act on two time scales. The short-term time scale attracts attention, focusing conscious awareness, while the long-term time scale, trial by trial, forges synaptic strength and thereby influences learning. *Liking* involves opioid activity in the nucleus accumbens and results from evaluating objects and events by the pleasure they will bring. Humans and many nonhuman mammals behave in very similar ways in response to this opioid activity, producing strikingly similar facial reactions (like tongue protrusions) when presented with something they like.

How is this relevant to the biology of explanation? The answer to that question is somewhat complex. There are places in the brain (in the back of the parahippocampal cortex) that act as "association areas." These areas make connections with stored information, and when an experience we are having interacts with stored information ("the last time I had this experience it ended up feeling really good"), a special kind of receptor (so-called mu-opioid receptors) make those associations stronger, and the perceptual pleasure more intense. A good part of the pleasure of learning comes from the coherence of these association regions. Pleasant experiences can stand out from others only if our normal level of pleasant experience is relatively lower. This is achieved by special neurotransmitters called endomorphins (of which mu-opioids are one). They "calm the mind" by making synaptic connections with GABAergic neurons. GABA (gamma-aminobutyric acid) is associated with calming, even lethargy, and provides the path for quieting the trembling of Huntington's disease patients. GABAergic neurons generally inhibit neuronal firing. Their release of GABA has an inhibitory effect on much of the brain most of the time, eventually allowing for greater stimulation.

So mu-opioids in this case act like "uppers," causing you to be more alert and attentive and stimulated, which not only contributes to the feeling of pleasure but also pushes you to do more of this "perceiving" thing. There is also an interaction with the dopamine system—opioids inhibit the inhibitors, allowing dopamine to do its reward-training job.

In Chapter 4, we will explore how all explanations depend on Inference to the Best Explanation (IBE). We will see that IBE is a cyclical reasoning strategy. Learning, too, deploys that same cyclical structure. And it is no mystery why we learn from IBE, because our learning cycle has the structure of an IBE. Although no primer on the neuroscience of wanting and liking will convey the complexity of these states of desire and affect, this compressed story does illustrate how desire can be at the bottom of the learning cycle. We are biologically disposed, so to speak, to get a good feeling out of learning.

The independence of wanting and liking, the fact that they are separate neurological entities, helps us explain so many of the oddities of desire, since we often think of desire as a *combination* of liking and wanting. We can really want to gamble or shop even though we don't expect to gain that much pleasure from it, making it an irrational, compulsive desire that prompts impulsive behavior. Drug-addicted desire is another good example of irrational wanting, in which the drug addict often knows that the drug's full impact is negative but nevertheless desires the fix, even though he or she may also know that it will only bring cessation. Aesthetic appreciation, on the other hand, is a state of high liking and low wanting, the experience of really liking Wyeth's "Christina's World" or Bosch's "Garden of Earthly Delights," but not wanting either displayed in your living room. This kind of state is sometimes referred to as subrational because it is an *enjoyed* stimulus—it brings pleasure—but it is not *craved*. The reasons we like it are not practical or "utilitarian." If you have low levels of wanting *and* liking, you would simply lack motivation, which is displayed to you and others as a feeling of indifference. This picture is unfamiliar to the person on the street, where to like is to want, and to want is to like (or expect pleasure). But this distinction allows you to generate a matrix of hypotheses, as shown in the following table.

LIKING AND WANTING KNOWLEDGE

		Feelings of Wanting (caused by mesolimbic area dopaminergic systems)	
		HIGH	**LOW**
Feelings of Liking (caused by nucleus accumbens opioid systems)	**LOW**	Feeling of Compulsive Need (morbid curiosity)	Feeling of Disinterest (boredom or indifference)
	HIGH	Feeling of Deprivation (intense need for cognitive closure)	Feeling of Interest (curiosity or aesthetic pleasure)

Adapted from Tables 1 and 2 in Litman (2005).

A potent combination lurks. Intense wanting and strong liking produce feelings of great pleasure, even uncomfortably so. A ravenous appetite for the objects of our desire leads us to expect the most delicious pleasure from satisfying it. The promise of this pleasure is what stokes our motivation. Like two streams meeting, "wanting" and "liking" together work to reinforce learning through rewards of different kinds. Because dopamine neurons can predict rewards, dopamine is a central element in reinforcement, motivation, and learning. By associating the activity of explanation with the feeling of pleasure, it becomes a learned behavior.

The lessons of this research on curiosity and desire for explanation should be clear. Explanation, like all human cognitive processes, is grounded in our biology. It is not just the psychology of understanding that drives the seeking for, and acceptance of, explanations: it is also the biology of motivation, the visceral arousal of urges to explore. We *want* to learn, and we also *like* it when we do. When we pause to marvel at the accomplishments of modern science, we acknowledge not the elegance of a formal computational process, nor an aggressively reconstructed explanatory sequence designed to suit the taste for precision and order so valued among many philosophers of explanation. Instead, we see something much less formal. The description of nature as irreducibly statistical was so profoundly unsatisfying to Einstein, for example, that he sought relief in an image that "made sense" to him.

While scientists have special training in the arcane details of their chosen subject, they have no special training in explanation. When it comes to the activity of explanation, scientists are just as good—or bad—as everyone else. Yet, with only the explanatory tools acquired at their parents' knee, scientists deliver accurate descriptions of the world that range from the mundane to the majestic. This kind of account of Acceptance of Explanation is less dignified than we may have wanted. It is designed to accommodate the uneven performance of explanation and the routine distortions that sometimes shape our embrace. As we will see in cases important in public choice, people can have a positive attitude toward false beliefs—including false theories—even when they have credible reason to think they are false. And this positive attitude disables critical scrutiny, giving the belief a kind of fluency. A number of empirical studies show that repetition of a statement produces processing fluency, and that processing fluency creates a feeling of understanding, whether or not that statement is true.[8] The entrenchment of theoretical concepts and vocabulary, a staple of Kuhn's account of the dominance of a paradigm, can produce just such repetition-related feelings of fluency. Such influences on acceptance, as we have seen, needn't be truth-related. Indeed, sometimes fluency is produced by sheer dogma, as we find in the hydraulic, humoral treatments recommended in the textbooks of medieval medicine. Whether we want or like the objects of these theoretical desires for truth, consistency, or coherence, it is these desires that drive the learning cycle that then drives science. And these are the psychological, all-too-human forces that have dominated theory choice.

THAT PECULIAR FEELING OF UNEASE

Wanting and liking, being separate phenomena, can vary in their relative degrees of activation. The drug addict, for example, might show very high wanting for their fix but low liking. And when it comes to knowledge, curiosity also has the potential for very different trajectories of liking and wanting. Given that, and if we can also agree that the sense of understanding feels good, it is interesting to note, though not necessarily surprising, that not everyone pursues it equally. Just as there are different levels of interest in art, sex, music, and chocolate, there are different levels of interest in the sense of understanding. Differing amounts of curiosity is one such example: it is easy to think that everyone is curious about events that deserve it. Why do stars twinkle? Why doesn't the stomach, naturally filled with acid, dissolve? Each of these questions has an answer, but to find it, you have to wonder what it is; you have to be curious.

What, then, is curiosity? Classic conceptions associate curiosity with satisfaction and pleasure. Ruskin claimed, "Curiosity is a gift, a capacity of pleasure in knowing."[9] Samuel Johnson offers a similar description: "The gratification of curiosity rather frees us from uneasiness than confers pleasure; we are more pained by ignorance than delighted by instruction."[10] To cognitive psychologists, curiosity is a kind of desire. It motivates us to explore in order to learn. By this definition, curiosity is "a desire to know, to see, or to experience that motivates exploratory behavior directed towards the acquisition of new information."[11] The satisfaction of curiosity is powerfully rewarding. But like porridge, curiosity doesn't bring pleasure if it is too hot or too cold. If our curiosity is too great, it feels desperate, its object too out of reach; too low, and its objects don't engage our attention enough to reach the goal of understanding or insight.

It is easy, and probably not too inaccurate, to understand curiosity, or the impulse to explore, as a two-step process within the brain. The feeling of curiosity comes from the activity of the dopaminergic system, and the resulting feeling of pleasure from opioid stimulation. Feelings of curiosity have the tinge of anticipation, of a mildly tense sense of undischarged arousal. Pleasure is the feeling of that discharge, the satisfaction of an appetite gratified.

Litman proposes an "interest/deprivation" model of curiosity that combines both approaches, where curiosity is either a feeling of deprivation (CFD) or a feeling of interest (CFI), just as desire to consume food can be stimulated by nutritional deficits or a pleasing smell, but the pleasure remains the same in either case.[12] Both CFI and CFD turn out to be associated with self-reports of curiosity. And now that we have separated liking and wanting, the curiosity behind explaining is suddenly more complex. CFI, exercised in activities like contemplation of artistic work, is a kind of aesthetic state of high liking and low wanting. CFD is a state of both high wanting and high liking.

But the sense of understanding stemming from curiosity followed by explanation brings perils from the world beyond the brain. The biological story of the sense of understanding shows how many things have to line up to arrive at an accurate explanation, and still more for the sense of understanding to be a reliable cue of that correctness. Without a good theory, there is little hope of providing a good explanation.

CONCLUSION

Although the realism of this book may be decidedly non-Humean, this chapter is powered by Hume's famous (and doubtlessly correct) observation: reason alone does not move man to action. People are awash in motives for action. They may crave risk, hunger for competition, or revel in comprehension. In each case, it is a feeling that motivates them, a pleasant feeling of promised reward. And they often believe what they must to keep the pleasure coming. The feeling of understanding—which ranges from a rush of delight to a steady drip of satisfaction—is one such pleasure. It is what keeps scientists experimenting and theorizing when all seems lost.

No matter how powerful or persistent this feeling, it is not reliable. We regularly have the feeling in the presence of false beliefs and lack it when we believe truly. And yet, scientists are guided by it and, as a way of generating knowledge, the modern tradition traceable to Newtonian physics has no peers. We are yet to have a decent explanation for this striking success. But we do know the forces that have been obstacles to it—for example, commitment to false theories and poor heuristics or reasoning strategies. We navigate our work-a-day world so successfully because for simple beliefs in common-sense settings, we can use our sense of understanding as a reliable guide to truth. But as laypeople, we know little of gravitational lenses, polymers, stem cells, and cochlear innervation. Scientific knowledge is arcane, and so is the province of specialists. But now we are in delicate territory, where elaborate and intricate connections between the human mind and tendrils of the remotest corners of the universe cannot be comprehended by the crude heuristic—this sense of understanding. Overcoming this deficiency demands that we develop a prosthetic that tells us not just that our beliefs are true or false, but in what ways and to what degree. Scientific explanation and theory construction are those prosthetics. Physics, chemistry, biology, and some areas of psychology are known for their contributions not just to technology, but to the construction of elaborate descriptions of the unobservable world.

Once we take seriously the biological basis of the explanatory feeling, inference to the best explanation looks more like an informal narrative than a formal rule.[13] The learning cycle, after all, has the same kind of structure as the abductive argument for scientific realism. When scientists believe an explanation is true because it "makes sense" to them, it may be that the lucky guess produces a good feeling, and it may happen to be a true explanatory belief that is reinforced by the good feeling conveyed. In particular, if we trade pleasure or satisfaction for success, we have a close analogy. A behavior followed by satisfaction is more likely to recur in the future. While originally a behaviorist principle of operant conditioning, its newer applications manipulate subtle ensembles on unobservable neural causes. The cells of the nucleus accumbens fix the addict's behavior into a hopeless cycle. As we saw when discussing the biology of explanation, our behavior can be shaped or selected by its consequences.

Of course, other forces shape behavior as well, by determining the range and type of experience that the person can come to enjoy. A person's natural endowments may make some behaviors more enjoyable than others. Likewise, a person's cultural environment will craft pleasures by shaping tastes; a child raised in India is more likely than one raised in Denmark to dine on spicy Thali, in which the eating of it is reinforced by the pleasure it delivers.

The law of effect is not a speculative principle in evolutionary theory. There, the principle that favors successful (where success is measured by fitness) effects is natural selection. Instead, the law of effect operates just over (parts of) the course of an individual lifetime. Like evolutionary functions whose effects both explain and are explained by their causes, so the persistent behavior explains and is explained by the pleasure it brings. Explanations invoking the law of effect, like evolutionary explanations, have a cyclical character. As we will see in the next chapter, so does inference to the best explanation. That organisms have the tendency to leverage pleasure in this way is an evolutionary phenomenon. Which behaviors may come to be fixed by our satisfactions are matters for individual biology and psychology.

4

BELIEVING THE BEST EXPLANATION

WHEN STANDING BEFORE THE INTRICATE edifice of the scientific achievements of modern medicine, solid state physics, intergalactic astronomy, or molecular chemistry, you'd think it natural to have a sense of awe and wonder at the ingenuity of humanity and the power of our minds. And yet, the dominant theory of scientific progress removes the magic from these achievements: the slow and steady experimental method is rewarded for winning the race, incremental step by incremental step. However, scientific progress continues, and our method-dependent explanations of it have not kept up. If it turns out that scientific progress were the result of contingencies, like theoretical hunches, chance brushes with brilliance, fortunate geographic distributions, and unfortunate vulnerabilities to disease, then the marvels of modern science *should* be very surprising. And in fact, when we look closer at how scientific achievement has been unlocked, its explanation is quite jarring.

For those who champion the experimental method as the reason for scientific progress, successful explanation boils down to using the right steps in the right order with any phenomenon under study. Unfortunately, explanation as we find it is much messier. When we accept an explanation of a phenomenon, it is usually because a theory provides the best available explanation of that phenomenon—readily available examples include natural selection, regional stress, or rapid spread of disease. Peter Railton makes this point in saying, "Theories broadly conceived, complete with fundamental notions about how nature works—corpuscularism, action-at-a-distance theory, ether theory, atomic theory, elementary particle theory, the hoped-for unified field theory, etc.—not laws alone, are the touchstone in explanation."[1] It is hard to find formal constraints on good explanation that are both substantial and widely accepted. To take just two examples, predictability (from explanatory information) seems strained when applied to complex populations (say, of cells, for an individual's health) and historical domains (like evolutionary theory). And a noncircularity constraint merely rules out explanations that are trivial. In the absence of such formal constraints, we are left with rough and casual guidelines. And we can't hope to use broad, impressionistic tools of theory assessment to make a nuanced evaluation of a theory's overall promise, so our global assessment of theory quality is notoriously informal.

Progress in science, in fact, is based in induction. All induction, including all that is used by science, is abductive; it assumes or asserts that some observations support a particular law or theory because their doing so provides the best explanation for those observations. Denied this pattern of inference, it would be difficult to account for the way people reason inductively. Recall the mouse behind the baseboard: perhaps we see little droppings in the pantry and we infer that there is a mouse in the pantry, because this fact provides the best explanation of all of the evidence. Or we observe that a container's sides bulge with increases in interior temperature and infer the molecular/kinetic nature of gas because it provides the best explanation for this observed effect. In both of these cases, the investigators were looking for the best explanation. How do we decide on the best explanation? By turning to the informal assessment just mentioned. All induction is grounded on explanatory relations, right down to the straight rule, which says that objective relative frequencies should be interpreted as limiting relative frequencies, on the grounds that the former converge on the latter. The straight rule has much in common with the Law of Large Numbers, which states that the observed distribution of a sequence of independent and identically distributed trials converges on the true distribution. We will develop this point more fully, but at the moment we will simply observe that no one applies these rules without supposing (either explicitly or tacitly) that there is an unobserved, or unobservable, structure that makes those rules true in a wide variety of settings. Without these principles, we would be unable to adequately explain the reliability of even the simplest inductive reasoning. My claim is that, quite explicitly, these principles play a role in the best explanation for the success or reliability of induction.

As we will see, a committed band of philosophers suggest that Inference to the Best Explanation (IBE) is an exotic brand of induction, in need of special defense. Admittedly, some inductive rules seem too simple to rely on explanatory principles. Take again the straight rule, and Hans Reichenbach's endorsement of it in *Experience and Prediction*:

> (SR) If you observe m apples and n of them are green, you should infer that n/m of all apples are green.[2]

This kind of rule looks so simple you might imagine it is foundational, and that its success does not depend on any further commitments about what nature is like. But it does. Projecting this rule into the future assumes that nature is stable in ways that matter to future judgments—that the sample is representative of the population, and that at least for the term of this projection, it will continue to be. When reasoning inductively, we infer from the sample to the population. But that is a choice that we make, not a fact about the nature of explanation. The fact that the ratio of the green to total apples sampled approximates the proportion in the overall population is part of the best explanation for the belief that there are n/m apples in the population, whether or not we choose to spend the time explaining the reliability of this assumption of convergence. So explanatory reasoning *is inherent in* inductive reasoning.

But abductive exceptionalism—the view that there is something special and additional about IBE beyond ordinary explanation—haunts the philosophical analysis of science. Philosophers have argued that some scientific theories, like those of evolutionary theory or historical geology, use abductive reasoning strategies and, because they do so, require a special kind of defense. They fail to see that all explanation relies on background theories. In a moment I will have the opportunity to bat away arguments from those who like their inductive principles fully denatured. But for now, the point deserves unshaded prose: all induction is explanatory, all of the time. There is no way to account for the reliability of even the most austere inductive rules without making explanatory assumptions about the nature of the population to which the rules apply. And if it ever appears otherwise, it is because scientists often ignore or take for granted much of what they take as background or standing conditions.

In this chapter, I will argue that scientific realism invokes a novel explanation for the theoretical success of science: IBE is an essential ingredient in all the inductive inferences so crucial to the explanatory success of modern science. IBE is prevalent in everyday explanation and, as it turns out, is just as common in scientific explanation, even if it is fully obscured by the more prominent features of inductive inference. IBE, dreadfully unsuccessful for false theories like hydraulic theories of medicine, is an intellectual treasure when reasoning about good theories like Newton's theory of motion or Crick and Watson's DNA theory of heritable properties. Critics will charge that this argument is a circular appeal to good theories and best explanations. However, the appeal is not about logical but causal structure. Good theories accurately track the causal structure of the world, and IBE exploits rather than constitutes that fact by appealing to multiple, independent features of the world. Although the descriptive evidence of IBE's importance and success is difficult to deny, many philosophers of science either ignore or argue against the explanatory structure of induction. By responding to these critics of IBE in scientific explanation, I will show how the history of science is not a procession of slow and steady incremental steps via the experimental method, easily undone by a steady flow of countervailing observations. Instead, under the influence of a good theory and this newly understood tool of induction, the explanatory success of modern science is like a one-way ratchet, ever tightening its grip as it advances.

So that we may see how IBE fits into a theory of scientific realism, it would be helpful to begin with perhaps the most influential form of scientific realism. Richard Boyd invites us to consider the doctrine of scientific realism "as embodying four central theses":

(i) "Theoretical terms" in scientific theories (i.e., nonobservational terms) should be thought of as putatively referring expressions; scientific theories should be interpreted "realistically."

(ii) Scientific theories, interpreted realistically, are confirmable *and in fact often confirmed* as approximately true by ordinary evidence interpreted in accordance with ordinary methodological standards.

(iii) The historical progress of mature sciences is largely a matter of successively more accurate approximations to the truth about both observable and unobservable phenomena. Later theories typically build upon the (observational and theoretical) knowledge embodied in previous theories.

(iv) The reality which scientific theories describe is largely independent of our thoughts or theoretical commitments.[3]

Each thesis has a role to play in what is thought to be the best explanation for the instrumental reliability of scientific methodology. But taken together, the truth of these theses constitutes the best explanation of the reliability of scientific methodology.

This robust version of scientific realism is understood to apply specifically to the mature sciences—certainly physics and chemistry, and to selected areas of biology. The successes here have been pretty uncontroversial (the theory of magnetism, the theory of gases, genetics, etc.), and their explanatory powers are conceded even among those who doubt that realism is required to explain these successes.

Now, turning to IBE, let's look at Peter Lipton's straightforward description, though this is merely one account of many:

> According to Inference to the Best Explanation, our inferential practices are governed by explanatory considerations. Given our data and our background beliefs, we infer what would, if true, provide the best of the competing explanations we can generate of those data (so long as the best is good enough for us to make any inference at all).[4]

The operation of, and reliance on, background principles suggests the close connection between induction and explanation. When asked to account for the reliability of the inductive inference from the blackness of crows here and now to distant or future crows,[5] we can explain the sustaining effects of genetic stability and environmental context. Induction by enumeration has this feature, as does every application of the straight rule. According to this view, some scientific hypotheses get an epistemic boost because they can explain. As Clark Glymour put it:

> One way to argue for a theory is to show that it provides a good explanation of a body of phenomena and, indeed, that it provides a better explanation than does any available alternate theory. The pattern of argument is not bounded by time or by subject matter. One can find such arguments in sociology, in psychometrics, in chemistry and astronomy, in the time of Copernicus, and in the most recent of our scientific journals. The goodness of explanations is

> a ubiquitous criterion; in every scientific subject it forms one of the principal standards by which we decide what to believe. A philosophical understanding of science should, therefore, give us an account of what explanations are and of why they are valued, but, most importantly, it should also provide us with clear and plausible criteria for comparing the goodness of explanations.[6]

Two importantly related points must be made about IBE. First, even in schematic form, it is clear that straight rule arguments like "If you observe m apples and n of them are green, you should infer that n/m of all apples are green" still carries risk. Circularity charges against inductive arguments as a category, then, seem misplaced—after all, invoking the principle that a sample is relevantly like a population is unsuspicious; indeed, the principle is heavily worked and widely reliable. So while some philosophers of science may indict distinctly philosophical arguments for realism, the shame borne by these arguments can't come from their reliance on induction. As an inductive principle, the general reliability of sampling is not one "whose validity is itself under debate."[7] Second, some instances of this argument form are demonstrably bad, and no one who defends such explanatory inductive arguments holds that it is always a truth-transmitting inference.[8] For example, 16th- to 17th-century theories of digestion were poor: the Paracelsian theory postulated the existence of the stomach as a "cold and dry" organ, while da Vinci hypothesized that the function of the stomach was to improve respiration. Van Helmont at least treated the stomach like a chemical chamber, but this image came more from the mysteries of alchemical affinities than the verities of modern chemical combination. In this environment, the winner, the best available theory, may be nothing more than the devil you know. These two points are related because it is the theory-dependent and inductive, ampliative character of abduction that makes poor conclusions possible. Of course, this raises the question of what makes for a good inference.

For example, are there any relevant differences between an inductive inference that the sun will rise tomorrow (based on a count of the many past days of sun risings) and an inductive inference that evolution is true (based on a count of the many observations in geology, botany, and zoology)? The former induction seems like the result of an innocent count, while the latter appears to be far more complex, and thus less likely to be valid. Yet the reality is that cases of induction don't differ in *whether* they invoke explanatory principles, but in the *extent to which* they do. The reliability of the first inference is mysterious without a very general explanatory principle of the uniformity of nature, while the second induction merely *appears* mysterious without a more specific, explanatory principle of natural selection.

Modest and aggressive inferences alike invoke explanatory considerations. Even inductive reasoning that appears to be merely enumerative is, at bottom, explanatory reasoning. Take the archetypically enumerative marbles in an urn. The fact that differently colored marbles drawn blindly from an urn (and replaced) will converge on the actual distribution of colors in the urn is explained by principles like durability of colors and the independence of marble color from marble position in the urn. The fact that we abbreviate explanations by tacitly accepting structural assumptions about population parameters, taking them for granted, doesn't make those assumptions any less explanatory.[9] In his book *Tychomancy*, Michael Strevens cites lots of surprisingly complex background assumptions in many fields that are "built into" our inductive reasoning. To take just one example, Strevens examines the assumption that historians tend to make about the distribution of injuries in the Battle of Sommes. It is clear that people assume that the distribution of bullet injuries on the soldiers is random—and it is. People seem to reason, in a way that is deeply implicit, that the random pattern of variation follows from the limited motor controls of the soldier shooting the gun. The requirements to make the distribution nonrandom (say, mainly in the vital organs) far outstrip the precision of the instruments—the gun barrel, site, and soldiers' motor functions in holding steady aim, for example.[10]

By itself, mere description is not of much use in explanatory science; it only tells you things like the distribution of properties in a population—the ratio of brown to non-brown eyes in a human population, for example—not *why* the property is distributed in the population in that way. For that, you need to appeal to the operation of unobserved or unobservable properties like genetic dispositions that cause that distribution. Only then can you *explain* why the population has the current distribution. In science, if we want to move beyond mere description, we will always appeal to unobserved or unobservable mechanisms or processes.

Inductive inference is successful because populations and the samples drawn from them share common structures, and this overlap allows induction to generalize from a sample to a population. The statistical concepts and principles that depend for their justification on the population's structure are, for example, the independence of random error, statistical power, and homogeneity of variance. Because of this, all inductive reasoning is explanatory reasoning—it either explains the phenomenon in question or it plays a supporting role in explaining it.[11] That's because the strength of every inductive argument depends on explanatory principles ranging from the modest to the substantial, from "the future will in relevant respects be like the past" to "as the observations increase, the sample mean will tend to converge on the population mean."

In the context of scientific realism, there is no reason to suppose that a bad hypothesis couldn't be *good enough* to make *some* inference or other, no matter how qualified and modest. Conceptions of IBE such as Lipton's can permit that a bad hypothesis can lead to a good inference so long as the bad hypothesis is still the best of the lot and is at least good enough to make any inference at all. As Quine puts it:

> New groupings, hypothetically adopted at the suggestion of a growing theory, prove favorable to inductions and so become "entrenched." We newly establish the projectibility of some predicate, to our satisfaction, by successfully trying to project it. In induction nothing succeeds like success.[12]

But notice that, once entrenched, it doesn't matter where those hypotheses truly came from. In fact, they could be drawn from theories arrived at by accident. This thesis of accidental origin could be combined with the "No Miracles" argument for scientific realism, the argument that the success of science would be miraculous if scientific theories were not at least approximately true descriptions of the world.

Therefore, realism does not require the elevation of method. In fact, in fields where there are nothing but poor theories, all the methods in the world, including the highly lauded and exalted experimental method, may be insufficient to improve the theory. A great cook, following the best methods, will produce a lousy meal if the ingredients are bad. Scientists' inductive practices are reliable routes to the truth only if the theories they are using are approximately true in some relevant respects. Assume that out of method alone can come truth, and you have little explanation of why theoretical science stagnated for centuries from the inception of robust experimental methods in Al-Kindi's 9th-century Muslim world and of Grosseteste's 13th-century world of England and the Northern European continent, and why it continues to stall in some fields. On the other hand, if you have a good theory, even by way of grand and dramatic hunch, method may still only play a subsidiary role in securing progress. To continue the analogy, with amazing ingredients, even a poor cook can make a nice omelette. (The methods can't be awful. But you can make a lot of mistakes and still produce a pretty tasty meal.) For example, while phrenology had a mystical origin and an inglorious end, the modern version of cerebral localization grew from phrenology, and from a vivid bet that the localization of function postulated by phrenology could be confirmed by multiple and mounting sources of evidence from local muscle contractions in the cortex. Using newly developed electrodes, scientists like Hermann Munk had demonstrated in the 1870s the functional localization of vision in the occipital cortex.[13] The breakaway hypothesis of localized function for speech or for vision had been confirmed by the careful application of experimental method that never would have produced probative evidence based on much poorer theories that treated all parts of the cerebral cortex the same.

This form of scientific explanation can, in fact, be used to explain the success of science itself, and is intricately involved in explanations for the origins of modern science: the corpuscular conception that came out of late alchemy captured enough of the actual features of atoms and molecules, even while alchemy was a "bad" science, and did so accurately enough that this conception could spawn the unprecedented theoretical progress of the West.

Not surprisingly, using a reliable rule in reasoning can give you evidence of its reliability. For example, if "Infer the truth of the deductive consequences of the dominant theory" is a reliable inference rule, its reliability might also count as *evidence* of the approximate truth of the dominant theory. This is one thought behind a traditional argument for scientific realism. Accepted theories have to be good enough to identify plausible interfering artifacts and control for them. For example, the measuring instruments used to record signals from nerve cells have to be insulated from electrical house current. The argument next calls attention to the apparent reliability of this methodology, which, after all, has yielded, and continues to yield, impressively accurate theories. In particular, by relying on this methodology, scientists have for some time now been able to find ever more instrumentally adequate theories. From these descriptive facts, Richard Boyd then argues that the reliability of scientific methodology is best explained by the hypothesis that the theories on which it relies are at least approximately true.[14] From this and from the fact that these theories were mostly arrived at by abductive reasoning, he concludes that abduction must be a reliable rule of inference in the context of mature sciences.

The epistemic dependence of scientific progress on theory quality—and the consequent overemphasis of rules or canons of experimentation in improving a poor theory—is easy to illustrate. Remember the inference rule just discussed: believe the deductive consequences of the dominant theory. This rule will work very well if the dominant theory is an excellent one, like the current theory of optics or thermodynamics. But it will work poorly if the dominant theory is poor, like the extramission theory of vision or the humoral theory of health. The contingencies of history, psychology, and culture may deliver theories that differ in their success, but the rule is the same. And it is this insight that underlies some scientific realists' claim that an inductive practice like those suggested by IBE can be safely applied only once you have a mature theory.

Conversely, if our background theories are poor enough, it is possible that the best explanation will be pretty bad. Recall the qualification that IBE works with a good theory, not with a bad theory. Between these extremes of good and bad theories, there is a lot of middle ground and no principled answer to the question of when a theory is "good enough." For example, Newton had a hunch that the world in the small was like the world in the large. He wasn't the first to believe in atoms: he was preceded by philosophers of antiquity and fellow scientists in the emerging academies. He wasn't even the first to formalize a corpuscular hypothesis. There was Boyle, Hooke, and Locke, his closest predecessors, and Galileo before them.[15] But Newton was the first to develop laws of motion that applied to molecule and missile alike. Both have velocities predictable from Newtonian forces, but for most purposes we can ignore gravitational influences on molecules, but not missiles, when we make those predictions. Until this "good enough" theory came along, however, IBE about atoms did not make much progress. But Newton's unifying pattern provides the basis of a powerful inference to the best explanation.

A critic of realism can always *say* that alternative, demonstrably inferior theories have equally admirable epistemic or pragmatic virtues compared to the current (preferred) theory, but it is harder to show explicitly that the virtues in question are in fact genuine or significant. As Stanford puts it, "It was perfectly reasonable for nineteenth-century teleo-mechanists to regard the existence of developmental or formative vital forces as supported by multiple lines of independent evidence converging from diverse sources."[16] Well, on a certain conception of reasonableness, it might also be perfectly reasonable for medieval theorists of humoral medicine to regard their view as supported by multiple lines of independent evidence converging from diverse sources. But that's not saying much, because the independent evidence they regarded as converging provided only superficial and fragile integration of theories—and terrible advice for medical practice. It may have appeared that the tongue of a stroke victim was immobile from hydraulic tumescence rather than neural paralysis, but vivisecting the veins beneath was an awful treatment and had disastrous outcomes that were casually observable, making their view both incomplete and deeply mistaken. Having better theories, by whatever means—hunches, psychological peculiarity, unreliable extension of a key concept from another domain, the opening of new trade routes, and so forth—would have vastly improved the accuracy of one's belief that some theoretical commitment is supported by diverse lines of independent evidence. After all, reliability demands only that the process *be* accurate, not that the process allow you to predict a theory's accuracy.

Of course, if a realist were to make an argument in the following form, it would be at best question begging: the abductive argument for realism is reliable because abduction is reliable (or at least not prima facie unreliable). But this is not the realist response; the realist would treat this argument as enthymematic. Once the argument is properly spelled out, the realist is able to make use of a variety of resources, each of which has in its favor an independent argument, in the same way that the apparent circularity of "Sleeping pills work due to their dormative powers" goes away once you describe the (independent) chemistry that accounts for the dormative effects. Granted, not all evidence is equally direct. But it does accumulate, and if you did not explicitly depend on substantial inductive knowledge, you would be denying yourself significant sources of confirmation. Recall that "There is a mouse behind the baseboard" best explains the evidence of a scratching noise, droppings, and gnawed food. Without each piece of evidence, the hypothesis is less well confirmed. But let's graduate to a richer theoretical inference. Evidence that evolution best explains the diversity of species requires a subtle web of evidence inaccessible to direct observation. Instead, we infer the causes from peripheral traces in the fossil, geological, and astronomical record.

To make matters even more complicated, part of what we know about the fossil record comes from the physics and chemistry of radiometric dating, and part of what we know about the geological record comes from what we know in astronomy about the composition of asteroids and the timing of impacts. Because all of the evidence is joined by the sinews of nature, any revision of a piece of this evidence would require, quite implausibly, the rewriting of vast bodies of existing findings. This brings us to a striking consequence: evolution explains the diversity of species, and the diversity of species also explains evolution. How is the diversity of species supposed to both explain and be explained by the theory of evolution? This sounds circular. And admittedly, there is an apparent circularity when inferring the truth of a claim from the fact that it best explains some set of facts. However, after going beyond the superficiality of the apparent circularity there, the explanation becomes quite elegant.

To take another example: How do we know that atomic clocks are accurate (more accurate than a spring clock)? Because atomic theory tells us so. And how do we know what atomic theory tells us is true? Because it provides the best explanation for a wide range of phenomena, including the accuracy of atomic clocks. Once again, we have an appearance of circularity, but it is abbreviation that creates the appearance of circularity. After we walk through the details of atomic theory related to atomic clock making, notice how much less paradoxical and circular the explanation seems.

A standard analog clock (like a mainspring watch) and an atomic clock both make use of oscillation to track the passage of time: a standard clock keeps track of passing time because the balance wheel has a regular period of oscillation, while an atomic clock is a clock that uses the resonance frequencies of atoms. That is, an atomic clock keeps regular time by exploiting (or tracking) the oscillation between the atom's nucleus and the surrounding electrons. The "spring" in an atomic clock is the oscillation frequencies within the atom. This provides for a highly refined timing mechanism: the mass of the nucleus combined with the gravity and electrostatic elasticity (or resilience) between the positive charge on the nucleus and the electron cloud surrounding it. It is not subject to mechanical warp or wear over time.[17]

Not only are atoms impervious to mechanical warp or wear, but also they have *exceptionally* consistent resonating frequencies. All Cesium-133 atoms, for example, always resonate at exactly the same frequency: 9,192,631,770 cycles per second. This consistency has a specific physical basis: an energy change of an atom or molecule produces an emission or absorption of microwave electromagnetic radiation. The oscillation is controlled by the frequency of that production.

Depart in any substantial way from this explanation, and not only does the explanation for the accuracy of atomic clocks fail, but also most of what we believe about processes at the atomic level fails. We can entertain a philosopher's highly qualified doubt about the commitments of atomic theory, but earnest assertions of doubt would force us to take action—to give up our beliefs that we can explain or understand anything that rests on atomic theory, phenomena like fission, fusion, and radioactive decay. Indeed, these theories' views are so tightly integrated that we cannot be selective in our skepticism. One effective doubt is the thread that unravels virtually everything we take ourselves to know about the unobservable physical world.

As with the atomic clock, many scientific hypotheses are routinely supported by the same observations they explain, but not *only* by the same observations they explain. The details are not just tied to each other. Typically, they are supported by a vast network of related facts, sometimes the core claims of several sciences. Modern atomic physics has allowed us to construct clocks of increasing accuracy (when compared to the old standard), and the new atomic clocks are then used to measure the duration of physical reactions. As long as the reliability of the hypotheses used in the construction of the atomic clock can be tested and established *independently* of those underlying the temporal measurement of physical reactions, a charge of circularity cannot be levied. The accuracy of an atomic clock appeals to different physical principles than those underlying the accuracy of a quartz clock.

So the best explanation for the accuracy of atomic clocks is that the oscillation is nearly perfectly repeatable. And the best evidence that these theoretical assumptions are true is that you can use them to construct a shockingly accurate clock. This bumper-sticker abbreviation of the accuracy of atomic clocks seems circular in just the way the offending arguments did for the reliability of vision and the accuracy of mature science. But like these arguments, the atomic clock argument contains too many independently testable principles that their establishment could not be documented on a bumper sticker. As Lipton notes, this "mutual support" or "apparent circularity" is common in science:

> where the phenomenon that is explained in turn provides an essential part of the reason for believing the explanation is correct. For example, a star's speed of recession explains why its characteristic spectrum is red-shifted by a specified amount, but the observed red-shift may be an essential part of the reason the astronomer has for believing that the star is receding at that speed. Self-evidencing explanations exhibit a curious circularity, but this circularity is benign. The recession is used to explain the red-shift and the red-shift is used to confirm the recession, yet the recession hypothesis may be both explanatory and well-supported. According to Inference to the Best Explanation, this is a common situation in science: hypotheses are supported by the very observations they are supposed to explain.[18]

The patterns of explanation set out here are anything but complicated or exotic. They are illuminating, but commonplace. Once we see that IBE is just good old inductive inference, the standard defenses of scientific realism are left standing, without falling prey to the criticisms of circularity, tautology, arbitrariness, and problematic imprecision. We are shooting for a standard of explanation that can be extracted from our best sciences, no more and no less, not one that has been rationally reconstructed by philosophers viewing the drama of science.

A persistent antirealist might say that we start to feel fluency about explanations that fit with more accurate *observational* theories, and so scientific realism gets no special confirmation from our reliance on fluency. But this is an antirealist boilerplate—ready-made for any theory interpreted realistically—not an argument against the detailed theoretical accomplishments spelled out in the atomic clock example.

Where philosophical distinctions and concepts have been useful *in* science, their merits are discussed and assessed. But IBE exceptionalism in induction (the idea that some inductive arguments, like IBE, have special and additional features) persists because of a rhetorical purpose it serves in the philosophy *of* science. This is disappointing; it is a field otherwise known for yielding to science when it conflicts with philosophical distinctions. Even so, philosophers and logicians routinely distinguish between inductive reasoning that is "merely enumerative" or "Baconian," in which you just count up confirming instances, and inductive reasoning that is "abductive." That principle would make the argument an example of IBE. Bas van Fraassen states that to meet the ideal standard of induction, IBE must be a rule, complete with a formal definition and distinctive conditions of application.[19] Larry Laudan faults realism for relying on so exotic a pattern of reasoning.[20] Arthur Fine endorses the charge that realism's success argument is circular, that the realist is "not free to assume the validity of a principle whose validity is itself under debate,"[21] and that the argument for realism advances "the very type of argument whose cogency is the question under discussion."[22]

Kyle Stanford combines all of these misgivings, saying that "sophisticated realists recognize the need for a *precise, nontautological, and nonarbitrary* characterization of the kind of maturity or special variety of success that is supposed to distinguish current theories from their superseded predecessors."[23] This sentiment is hard to dispute, yet it is misplaced. The question is how much imprecision is tolerable, not whether precision is even desirable. The history of science is littered with inappropriate expectations of precision and with hypotheses prematurely rejected for departing from wildly unrealistic standards of precision. Newton's own "Hypothesis non fingo" is an obvious one, in which he says he (and by extension, other science-minded people) can only know what can be observed, and so we must not make claims that go beyond what is observable. This is doubtless an inappropriately exacting general standard for science, and thankfully, Newton's practices violated his own injunction. Darwin couldn't explain heredity, but it would have been an enormous mistake to reject his theory on that basis. These cases of limitation and imprecision are not fatal weaknesses of theories, but simple descriptive conditions that theories and scientists face. The fact that there wasn't perfect clarity was inevitable (and likely even necessary) given that theories grow and mature. This conceptual imprecision isn't a bug, but a feature, of theory development.

At their most dramatic, realists argue that it would take a miracle for our theories to be as instrumentally reliable as they are and not be at least approximately true. As noted earlier, critics have accused scientific reliance on IBE of being circular. Peter Lipton also agreed with this concern about the inductive argument for the defense of realism, saying that "the scientific case for our best theories is what the No Miracles argument assumes, not what it establishes."[24] Antirealist critics contend that the argument rests on a premise—that scientific methodology is informed by approximately true background theories—which in turn rests on an inference to the best explanation for its plausibility. And the reliability of this type of inference is precisely what is at stake.[25]

For example, what is the best explanation for the success or instrumental reliability of the convergence rule in induction? The best explanation is that the rule is approximately true: as the number of items sampled increases, they increasingly approximate the distribution of the population. We can count up instances in which this principle reliably predicts the distribution in the population. In this case, each principle of justification is itself held to the standard of IBE. As a result, even the principle of IBE would be justified by IBE. This circularity, however, is not vicious, because its components are not foundational or axiomatic. IBE is not assumed from the start as an a priori principle but has the status of a defeasible, empirical principle whose reliability is an empirical hypothesis, one that is empirically confirmed in the process of seeking and formulating explanations. As long as IBE remains useful in explanations, it unproblematically satisfies its own requirements. And as it happens, IBE is more effective for theories in the more mature sciences.

Philosophers of science are sometimes seduced by stock philosophical arguments charging circularity in justifying scientific knowledge, and they use the comic book caricature traditionally used to establish skepticism or foundationalism.

1: My eyes are accurate detectors of visible objects.
2: What is your evidence that they are accurate?
1: I do well on vision tests.
2: What is your evidence that you do well on vision tests?
1: I can look at the results.
2: Doesn't that assume as a premise in your argument that your eyesight is accurate—the very conclusion you sought to prove?

Sometimes when people offer explanations, they try to follow pragmatic rules of communication, only offering information deemed relevant to the question asked. Often, we assume the best of our interlocutors, that they are smart and interested in advancing a case rather than winning an argument. This leads us to abbreviate explanations and leave tacit assumptions unobserved. But once you recognize you are in enemy territory, you are wise to offer a more elaborate explanation, spelling out the independent lines of evidence that indemnify the argument against charges of circularity. So let's pick it up where we left off:

2: What is your evidence that your vision is reliable?
1: The car that I see also makes a sound. The probability that I am prey to an error of both vision and audition is vanishingly small.

Aaaand . . . scene. This example dramatizes the fact that the reliability of our perceptual modalities are normally tested *diversely*, by what you see and what you hear, taste, feel, or smell. In fact, the reliability of vision can be tested diversely on visual evidence alone, from, for example, visual transduction (evidence from early-stage neurological or sensory information) and higher-level visual attention. Only a certain kind of object can produce arrays of visual information that can be integrated across such different visual mechanisms, and so the probability that unreliable mechanisms achieved this integration is tremendously small.

If vision is reliable, then the fact that you visually observed the results of the vision test *is* evidence that your vision is good. Of course, if your vision is unreliable, the appearance that your vision is accurate is uncertain. But the original complaint was not that scientific realists failed to answer the skeptic to the skeptic's satisfaction. Instead, this critic of scientific realism contended that you cannot appeal to the approximate truth of a mature science in an explanation of its success unless you gave reasons for thinking the theory was approximately true that were independent of its success (or its instrumental reliability). We may dramatize the application of this dialectic in the case of scientific realism:

Q: What is the best explanation for the success of science (or for the instrumental reliability of scientific methodology)?
A: That it is approximately true.

Q: What evidence do we have that our best theories are approximately true?
A: That they are successful, or instrumentally reliable.

Just as there is no magical measure that distinguishes tested principles that are dependent from background theories and those that are independent from them, there is no principled distinction between mature and immature sciences. When philosophers of science look at our best scientific theories, they have looked for explanations for their success. After all, we don't find this success just anywhere: just look at palm reading, astrology, or the theory of humors. Some believe that scientific theories deserving the most respect are mature—they are past what Richard Boyd called a "jumping off point." They are "both comprehensive and robust, i.e., supported by convergent lines of independent argument."[26] But others challenge this goal, arguing it is unreachable. And in the most famous expressions of this view, one commentator dismisses an expression of maturity as "tautologically empty."[27] But it needn't be empty. It might name the standing of a theory when it is good enough to reliably apply the inference rule.

As realists polish their defense of hypotheses drawn from our most mature theories, they introduce the idea that we use explanatory principles to support inductions that proceed to identify more remote or specific explanations. For example, once you have the general explanation of diversity in populations in terms of random variation, you can appeal to many of the same mechanisms or processes to explain the speciation that occurs in *isolated* populations. Clark Glymour states that these bootstrap principles

> are principles we use in our science to draw conclusions about the observable as well as about the unobservable. If such principles are abandoned tout court, the result will not

> be a simple scientific antirealism about the unobservable; it will be an unsimple skepticism.[28]

The sweeping consequences of this skepticism have been noted elsewhere. Richard Boyd replies to attacks on realist applications of abduction by pointing out that "the empiricist who rejects abductive inferences is probably unable to avoid—in any philosophically plausible way—the conclusion that the inductive inferences which scientists make about observables are unjustified."[29] If we want an explanation for the reliability of the statistical principles and concepts that underlie the sciences, we must suppose that we have at least modest theoretical knowledge. In particular, we must have knowledge that there is a dispositional structure that sustains the observed performance of the system[30]—an urn whose distribution of white and black marbles supports the observed distribution, a disposition that sustains the observation that tall parents have tall children, and a disposition that allows hybridized plants to be cultivated for traits we select. These dispositions exist, of course, even when the powers in question are not occurrent. In this regard, a poor theory actually *interferes* with the engagement of these underlying theoretical structures, mystifying the explanation for the black-to-white marble ratio, tall children, and a weeping cherry. We deny this modest theoretical knowledge at the cost of neglecting observational success.

Despite the logical overtones of a circularity charge, the most persistent charges from antirealists seem to result from explanatory demands placed on scientific realists that they are then claimed not to meet. We should assume that this explanatory expectation is what underlies Kyle Stanford's objection to explanationist defenses of realism from unconceived alternatives. There are, we might recall, a potentially infinite number of alternative theories that might have been selected for pursuit, any one of which may be "more right" than the theory we use.

Beneath Stanford's efforts is a peculiar explanatory standard. Although Stanford goes to great pains to show that his objection to a scientific realist explanation for the success of science from unconceived alternatives is different from the objection from the underdetermination of a theory by evidence, antirealists can get the objection going only if they impose a superogatory explanatory demand on explanations for the success of science. The fact is, at any juncture in the history of science, there are many unconceived alternatives that remain unpursued, some salient and some not. For this to be a problem for scientific realism as opposed to a description of a challenge it has overcome, the critic has to believe that the scientific realist has to explain or justify the choices—perhaps just the salient ones—that the scientists *didn't* make. The argument from unconceived alternatives comes dangerously close to entailing a specific explanatory standard to the effect that you must show that *none* of the unconceived alternatives were actually responsible for science's success, or that failing to demonstrate that responsibility exposes a special epistemic problem. But this standard is undesirable. We can understand how we arrived at a theory without looking back, predicting its every meander. To explain how the actual choices we made led to the success of science, we don't have to describe and justify every choice we didn't make. This would be not just an extravagant explanatory demand, but an exotic one as well, operating nowhere among practical scientists or those charged with explaining, and so understanding, science.

One might grasp for exoneration from this realist refutation by claiming that what holds *in* science doesn't automatically hold *about* science. But this seems a strange dichotomy for those focusing on science for its special epistemic properties. When it comes to standards of explanation *about* science, we can do no better than to follow the explanatory standards *in* science. Or at least, you should have to explain why you think you could do *better*, and *why*. If the defender of unconceived alternatives chooses now as the time to impose pretheoretic, distinctly philosophical standards of explanation, the apparent desire to illuminate explanations for the success of science is mere posturing by fair-weather champions of science.

We are all familiar with theories that once were thought for all the world to be true but turned out to disappoint. Ptolemaic astronomy, the luminiferous optical ether theory, the humoral theory of medicine, the vital force theory, and spontaneous generation are all theories defective enough that even some better theories are still not very good. But theories are, after all, complicated artifacts. There is a lot more to whether we should trust a particular IBE than the "maturity" of the theory from which it is drawn. Even the most mature theory can advance an explanation that will fail, because the precision implied in the explanation outstrips the theory's capacity to ratify it. The posits of string theory could easily be questioned, for example.

So when can we infer that the best explanation is also true? There is no magic (or even natural) criterion, but confirmation improves for the "best explanation" hypothesis as it unifies an ever larger number of disparate principles. In the end, however, progress doesn't depend on an explanation that is the best, but only one that is good enough.

An explanation that is the best of the available competing explanations may be good enough to produce a substantial theoretical advance. These two conditions—unification and sufficiency—make the requirements of IBE clearer, but clumsier and less memorable. It would be nice to have a crisp answer to the question of when the best available explanation is "sufficiently good" to infer its truth. But remember our earlier injunction against automatically supposing that realism is a predictive theory. We might be tempted to ask, but we shouldn't expect a general answer.

Philosophers like general answers to such questions, and who can blame them? But when nature doesn't allow it, you have to move on. There is little point in implying that realists are uncooperative by not providing answers that nature doesn't permit. Whether we explicitly recognize it or not, we are satisfied with answers to questions that fall short of perfect generality, or sometimes not even having answers at all. When does a tool become art? When does a mist become a cloud or an offspring count as a new species? If such unanswerable questions occur *in* science, we shouldn't require that all general questions be answerable *about* science. There is an unavoidable indeterminacy in classifying cases of IBE as reliable. Depending on the presence of other factors, an available hypothesis may be sufficiently good if it has been tested by different methods on different evidence. Surely it is better supported than one that isn't. But that won't always work. Alchemists often tried to confirm the presence of the elixir of life by using more than one method, by heating and titration. So it is not the diversity alone that makes for a good test. The theory itself has to be good enough for the diversity of methods to display robust convergence on a single value.

It might be claimed here that skepticism is recruited in the philosophy of science not as a genuine expression of anti-scientism, but as a methodological pose to press us to explore the nature of knowledge, justification, reason, success, and reliability ever more closely. Even if dismissible in practice, its purpose is to prod. And after all, being curious about the reliability of science doesn't entail a rejection of science or its achievements. But if van Fraassen, Fine, Laudan, and other critics of realism are going to hew to scientific agnosticism, or its stronger brethren, skepticism, then it should not be just an intellectual pose. Deny the theoretical knowledge or reliability of radioisotope dating, and no significant portion of modern chemistry survives. No modern chemistry, no theory of bonding or decay. No theory of bonding or decay, no fruits of modern physics. Or, to focus on the instrument: selectively deny an inference to the best explanation about the products of microscopic inquiry, and our theory of radiation must be fundamentally unreliable. No decent theory of radiation, no microwave, no X-ray, no laser surgery, no radio-frequency ablation of cardiac arrhythmias, and so on.

If you really don't believe that scientific knowledge claims have met your standards, then stop using them, paying for them, and relying on them. Instead, politely pass next time you are offered antibiotics, or smile with a "thank you, no" when offered chemotherapy for a lump that has grown. Favor instead a ritual incantation, or better, ignore it altogether. This rebuttal has an uncharitable tone. But so does the insinuation that our core theoretical beliefs in physics, chemistry, and biology lack warrant. Antirealists will say they can accept the practical successes of science without believing the relevant theory is true. And, while accepting chemotherapy would involve an inference to the best explanation, antirealists would insist that the inference does not appeal to the unobservable. It is worth noting that this is a very selective, and likely arbitrary, way to be skeptical. Economists, for example, often make use of IBE exclusively about observables, but there is nothing specific about unobservables that would lead you to believe IBE is reliable above the line of unaided perception and all of a sudden unreliable below it. So if it is an uncharitable suggestion that an antirealist reject chemotherapy, so is the pose that we should give modern scientific arguments no more presumptive favor than we do palm reading, or that there is no difference in the quality of accepted and abandoned views. Is it fair to dismiss humoral medicine in favor of modern variants? Wouldn't such a dismissal signal an insensitivity to the modern view's indebtedness to its humoral predecessors? The most compelling practical reason to reject antirealist squeamishness about mature theoretical knowledge is its compulsory, seeping consequences.

Why don't all philosophers of science find the seeping consequences of antirealism troubling? People get away with posing as epistemic agnostics or skeptics because the complexity of their targets means it is no simple feat to explain the theoretical marvel that is every pharmaceutical concoction, every kitchen implement, and every polymer we handle. Yet a science built of theories has never been less than the default view, and we are too used to enjoying the amenities of modern science to imagine, at every moment, what life would be like without them. And because we take them for granted, we can at the same time pretend to not care, to treat them like disposable curiosities.

If the realist is right—that good explanation rests on an accurate conception of causes—then it will be possible for good explanations to have no audience. This idea will be explored more fully in the next chapter, but to take just one example: a model of watershed pollution may contain a hundred variables, but the cognitive limit on working memory makes it impossible to appreciate even the first dozen factors indicated in the explanation, not to mention their respective powers and directions. The joint effects of these causes are not, and perhaps even *could not* be, understood by humans. Success can have a contingent cause, arriving decades before anyone recognizes its power, and reliability can be unarticulated. In this case, scientific progress may have none of the internal hallmarks of intellectual achievement, or none of the hallmarks, if you prefer, of understanding. In this light, we should understand the original charge of circularity in scientific realism as one not about a defective explanation, but about frustration with its incompleteness. In this respect, Lipton makes the feeling of circularity sound more troubling than it needs to be. We may not understand how a rule that warranted one inference could deny another. Or we might not understand how the approximate truth of a background theory could explain the success of a particular science. It is a frustration that our knowledge is limited. But a rule can still be reliable without us being able, practically or possibly, to say why.

Students of science are trained in a triumphalist vision: science is an incremental march to the step of the experimental method. Sadly, this is just not the case. Yet the vision of this chapter is still triumphalist, just not necessarily incremental. These lessons don't fit the dominant narrative of linear, piecemeal ascent. The power this narrative assigns to the role of the human mind fits nicely with the Enlightenment lesson that reason eventually bends the world to our will, or ultimately forces nature's secrets to the surface. While most people accept the idea that some views about nature are accurate and others not, that some are right and some are wrong, we are slow to embrace the idea that mature true views come to us by accident, as a contingent consequence of cultural identity, a scientist's personal idiosyncrasy, a quirk of training, or an economic need. Risking smugness, we might say that when it comes to progress through experiment, while it is important to be smart, it is at least as important to be right. If we equate subtle methods with smarts, and attachment to true beliefs as being right, then this is exactly the contrast most illuminating if we are to understand instrumental and theoretical success in the history of science.

For more than 30 years skeptical philosophers have charged that abductive inferences are unjustified, claiming you can infer the truth of a hypothesis only if you suppose that we can have knowledge of unobservable phenomena. The genesis of this argument goes at least as far back as Hume, who noted that even the secure inference that the sun would rise tomorrow could be consistently doubted, because it depended equally on the truth of an assertion about another unobservable state, that the world in the future would be, in the relevant respects, the same as the past. But Hume's point was that inductive arguments fall short of deductive certainty, not that they are especially suspect. If they were so unreliable, we would never be able to infer that there is a mouse behind the baseboard, a bunny behind the woodpile, or a single grid under all the microscopes. It's no secret that inductive reasoning falls short of certainty. But if you try to parlay that modest claim into the apocalyptic claim that assertions of theoretical progress in science are unjustified, you have officially turned a warning about complacent or self-satisfied trust in science into a lament that induction does not carry the certainty of logical demonstration.

If the experimental method were sufficient for theoretical progress, we would have found that theoretical progress in the Islamic world from 850 to 1200. If it were necessary for theoretical progress, it would have been more richly involved in the early days of Boyle and Newton. If a modern physics textbook must have a one-paragraph history of science, let's make it an accurate one: it's not that there was a good method; it's that the theories got better. That one-paragraph story may be unsettling, but at least it will be true.

Of course, everyone thinks their one-paragraph story is true, and we are too often wrong to be so unguarded in this embrace. Awareness of this fallibility leads to the pessimistic meta-induction: all scientific theories in the past have been false. Therefore, our current theories are likely to be false. This argument courts drama by ignoring the tremendous scientific advances in theory and labeling all of them false simply because their scope has been delimited, or they fall short of perfect accuracy. But in a world of nuclear bombs, molecular fans, genetic engineering, magnetic resonance imaging, and so forth, it is unduly cynical to dismiss these developments as no better than epicycle theory—observationally predictive but theoretically false. And it is needlessly despairing to throw all theories that fall beneath God's standard into one category of false intellectual projects. Modern chemistry is better than alchemy, and the modern physiological theory of axonal conduction is better than the early modern theory of animal energies. Some theories are truer than others. If there is a contentious sound to this argument, it is only because philosophers' penchant for deduction does not always dispose them to weigh the evidence. But for philosophers who weigh the evidence rather than lament departures from deductive certainty, the call is an easy one to make.

5

GOOD THEORIES, LUCKY HUNCHES

THE POWER OF CONTINGENCY AND THE ORIGIN OF SCIENCE IN THE WEST

Few people are so smitten with the human species that they believe we can understand any and every true explanation that can be formulated. There is an important psychological lesson in the fact that the human species has an "Umwelt,"[1] a particular and unique—and also limiting—way of experiencing its environment. To appreciate how humans could suffer such stark boundaries, we can examine the drastic limitations and spectacular skills of other animals. Just as true explanation is beyond the Umwelt of other animals, some true explanations are too complicated for humans to cultivate understanding.

For example, the earless reptile Tuatara lives in a different frequency and decibel range than the human. Even without ears or a developed inner ear, Tuatara can still show a frequency response from 100 to 800 Hz. Most of the human's world is silent to the Tuatara, who cuts a narrow 700-Hz swath through the human's spacious 20,000-Hz range.[2] The Tuatara can never have an understanding of sound as experienced by humans; that is beyond its Umwelt. Yet, we play the role of the Tuatara to many other animals. For example, elephants may feel seismic waves to locate subsurface water, waves not detectable by our sensory manifold, and so a skill utterly beyond us.[3] Using little buds on their lower lip called Lorenzini ampullae, dogfish sense their prey (such as flounder) by picking up a weak electric field that is imperceptible to humans.[4] Pigeons home by a complex set of geomagnetic, visual, and atmospheric scent cues.[5] None of these cues can be perceived by humans; they are not part of our Umwelt, and thus we simply can't experience these cues.

These species have highly skilled behaviors that they can neither explain nor understand, but we would never claim that they are not successful. Just because the dogfish can't name their Lorenzini ampullae or understand how they work to find prey doesn't mean we say that they don't work. Humans, too, are in this position. We may not understand our own situation. The contingency of the human's Umwelt, together with our unique adaptations to it, fixes the kinds of things that we can spontaneously explain from those that require time and study, and those that may go beyond our current ability to comprehend. Just as the dogfish's Umwelt means it is closed off from understanding how it hunts for prey, the human Umwelt may close us off from some true explanations. There is a difference between there being a right explanation and being able to comprehend it. Given the distinctive psychological *limitations* of different species, it is a matter of psychological or biological contingency which problems in their environments they can solve. In the case of humans, it is a question of which discoveries can be made.

Scientific progress must be achieved through this idiosyncratic filter of human experience. What are some examples of cases where an adequate explanatory model is unlikely to produce any understanding in anyone, due to computational complexity?

Complex as our genome is, it obviously can be comprehended by *something*: our cells do it every day. Or, as the physician and essayist Lewis Thomas once noted, his liver was much smarter than he was, and he would rather be asked to pilot a 747 jet 40,000 feet over Denver than to assume control of his liver. "Nothing would save me or my liver, if I were in charge,"[6] he wrote. In a similar vein, we may never understand the workings of our cells and genomes as comfortably and cockily as we understand the artifacts of our own design. "We have evolved to solve problems," Dr. Evelyn Fox Keller said. "Those do not include an understanding of the operation of our own systems—that doesn't have much evolutionary advantage."[7] It's quite possible, she said, that biology is "irreducibly complex,"[8] and not entirely accessible to rational analysis. Human beings are arguably the only animals capable of metacognition—the ability to represent our thoughts to ourselves, and to explicitly monitor our psychological processes. With this ability comes the consciousness of a dichotomy between a mental process and our awareness of that process.[9]

On the account of explanation I defend, what makes something a good explanation is a relatively simple affair: an accurate description of the underlying causes that bring about an effect. If this is correct, then there are many explanations that have nobody to understand them. In fact, many true evolutionary, biological, population-ecological, and cosmological explanations may be tracked by our cognitive capacities, perhaps using heuristics, that are not fully and accurately represented by the natural limitations on human attention and memory. Even so, despite our natural cognitive imperfections, there are powerful signs that we might be on the right track, signs of confluence or unification of theories. We know that these signs are reliable because we have documented their operation in populations with known values. Consider the most cherished of these signs: the agreement of quantitative values under diverse testing.

The value of diverse testing is clear. The hypothesis that two animals overlapped in evolutionary time is better confirmed when it comes from two largely independent sources—in this case, either by proximity to the same catastrophic event (like the Tioga ashfall) in a fossil bed or by radioisotope dating each fossil. So there are two distinct methods of measurement, using different instruments and units, converging on the same result—that the two species coexisted. Or consider the hypothesis that heat is kinetic motion and will increase pressure in an enclosed volume. This can be confirmed by either decreasing a container's volume or increasing the flame beneath it. When an object engages us in two or more ways, it is all the more difficult to dismiss its existence, because the probability that the measurements would agree but the object doesn't exist, or is very different in character, is vanishingly small. Each method introduces different kinds of measurement errors (assumed to be randomly distributed), and so the fact that the hypothesis survives under diverse tests is a testament to its robustness, the ultimate evidence of its existence. And there is no better example of a rationally negligible possibility than a contingent or accidental event.

When contingent facts like inclement weather, an unplanned meeting, or a serendipitous chemical mixture drive a scientific advance, the hypothesis receives especially potent confirmation, because the measuring techniques instrumental in securing confirmation couldn't possibly have conspired to produce the same outcome. After all, the introduction of different techniques was unplanned and unguided, perhaps even accidental. And it might be that the history of science is highly contingent, but that scientific success is all but inevitable. Perhaps scientific genius, human persistence, or the canons of experimental method virtually guarantee scientific success. Events could have turned out in any number of ways, but (almost) all of them involve great success in one form or another. So my thesis needs to identify events not simply where the history of science is contingent, but where its success itself is contingent. Success itself is contingent when a chance event made the difference between progress and stagnation or failure.

The hindsight bias prevents us from appreciating the power of contingency: when progress is undeniable, as it is in every area of developed science, it is easy to believe that it was also inevitable. In science, the sentiment proceeds, nothing is left to chance: project a hypothesis, manipulate a variable, record the outcome, and then repeat a slight variation. But this incremental image of the history of science obscures its most exciting features. The truth is, progress was driven by contingencies of creative talent, geographic location, social affiliation with the right people, the purposes of patronage, access to raw materials, and the needs of industry or the military. To spin the history of science into a story of rational development, you need an unscientific tolerance for inaccuracy, an aversion to uncertainty, and a compulsive need for narrative closure. The reality, however, is far messier. This view of history as contingent has precedent elsewhere: cultural history, unlike the history of science, is much more comfortable with assertions of accident and contingency, but they are still jarring and dramatic. For example, in his magisterial *Guns, Germs and Steel*, Jared Diamond asks why Europe emerged as the dominant power in the world. His answers often cite accidents—contingencies of history or geography—rather than the unfolding of a plan or policy. Military might is always an obvious cause of domination, because it is always the consequence of a carefully executed plan. But military might was not the whole story. For example, while the domestication of animals reaches back at least as far as Jericho, medieval Europeans were the first to live in close proximity to the animals they bred. This specifically European variation of domestication was, amazingly, a certain entrée to world domination: enemies, and potential enemies, were wiped out by the germs incubated in Europeans after the microbes made the jump from the animals they domesticated. And while it wasn't the sole cause of European dominance, it was a beefy contributor. European germs wiped out unresistant populations even before they could become enemies of the Europeans.[10]

If historical contingencies caused the rise of colonialism, why couldn't the same be true for the rise of science? But ask anyone on the street about how science originated, and if you get an answer it will likely be a fanciful story that begins with the experimental method, drab and uninspired. According to that narrative, only after the method's introduction could science be done in earnest. Francis Bacon

(1561–1626) formulated a handful of canons for experimentation that will grind a passable meal out of any intellectual ingredients.[11] The experimental story is soothing for a number of reasons. The view that advances are ground out by sheer earnestness and dutiful application of existing knowledge assures people that, no matter how imagination may fail us, there are still great discoveries to come. Progress requires no grand theoretical landscape. By simply and faithfully applying the canons of experimental method, we can produce better hybrid plants, create stronger bridges, design better sewage systems, and breed better mice.

The problem with this generalization about the origin of modern science is that, like most generalizations, it leaks. Rather badly. The experimental method preceded Bacon by at least 400 years, and by itself it had little power to transform a bad science into a good one. Explanations of such success that invoke the rise of the experimental method, though comforting, seem feeble and limited. So the answer must lie elsewhere.

The truth is, given the huge lurches of theoretical progress documented in the historical record, the only way to explain the spectacular success of modern science is to assign a major role to contingent or serendipitous discovery and advance. Once we give due weight to contingency in scientific progress, we can for the first time give a more accurate account of the history of science.

All historians of science acknowledge the influence of contingency when it suits their purpose. In fact, the contingency of scientific progress explains otherwise inexplicable findings. Thomas Kuhn emphasizes the contingency in every theoretical movement: "An apparently arbitrary element, compounded of a personal and historic accident, is always a formative ingredient of the beliefs espoused by a given scientific community at a given time."[12] When a theory lacks the fertile body of beliefs that press the advance, "it must be externally supplied, perhaps by a current metaphysics, by another science, or by personal and historical accident."[13]

Modern textbooks also do not ignore the creative aspect of science altogether, nor the routine leaps that we make in our ordinary theoretical investigations. One chemistry textbook notes simply that hypotheses are "derived from actual observation or from a 'spark of intuition.'"[14] A representative, topical biology textbook observes that, in addition to using both deductive and inductive reasoning, hypotheses may also be generated in other ways, namely, by "(1) intuition or imagination, (2) aesthetic preferences, (3) religious and philosophical ideas, (4) comparison and analogy with other processes, and (5) serendipity." That is, "hypotheses are formed by all sorts of logical and extralogical processes."[15] Similarly, a biology textbook notes that "there is considerable creativity within the boundaries of these investigative processes. Insights can result from accident, from sudden intuition, or from methodical research."[16] Yet these insights, apparently, are still within the boundaries of a method. As we will see, the role contingency plays in the advance of science is far greater than even Kuhn and the textbooks give it credit for.

The beginning of science itself benefited from this sort of fortuitous contingency. The very best practices of protoscience amounted to the routine reliance on alchemical adepts (many of whom were frauds),[17] instruments that were poorly made or applied to the wrong phenomena, theories of mystical or occult origin, theological conceit, and/or ideological imposition, all mistaken to their core. Science advanced despite its dependence on overwhelmingly false scientific theories of yore, the misleading feeling of fluency and understanding conveyed by bad explanations, the inefficiencies of human exploration, political and religious campaigns against science, and the world's complexity. Yet despite all of these obstacles, engineering exploits like bridge building, feats of practical medicine like the treatment of digestive ailments, and theoretical speculations about the underlying composition of the physical world bred a modern science that, together with democracy, must be counted as the greatest accomplishments of human civilization.

In the grand scheme of things, science happened virtually all at once. Human civilizations emerged in the last 15,000 years in Africa, Eurasia, and the Americas, and yet modern science was born and took shape in a brief period from 1640 to 1730, within a slender band of Europe. Why then? Why there? Why so successful?

For animals like us, it could be a crushing discovery that progress is often irreducibly fluky. It is disturbing, or at least a little unsettling, to recognize that one of your greatest achievements issued from sheer luck, from the spinning wheel of office mates, a casual conversation, or a weather pattern. Under an ideology that equates science with prediction and control, the role of luck or fortune seems incompatible with great discovery. We desire to find something out and, having gathered and evaluated the evidence, we have the feeling that the evidence is now under our cognitive control—that we understand. We try to fit all of the pieces of a process into a coherent explanation or unified picture. If we prefer to see ourselves as the custodians and beneficiaries of a just world, we need to tell ourselves that hard work is rewarded, success is not arbitrary, and competence comes from intellectual authority—we understand that which is under our cognitive control. But having looked at the biological underpinnings of explanation and understanding, we now sadly know that, in telling themselves a story of incremental and methodological progress, people create in themselves a feeling that they understand *that may or may not be accurate.*

After more than three centuries of post-Newtonian rhetoric about the bracing merits of doing experimental duty, it's time to examine the forces we can't control. Let's celebrate the importance of unplanned, unguided, and occasionally uninterested theorizing not as a dignified alternative to experimental reasoning, but as essential to it. But theory construction makes clear that real, lived science is chancy and haphazard, unpredictable and downright fluky.[18] This makes scientific progress historically contingent when the reasons for its occurrence were, from the point of view of some intelligible image of development, accidental.

CONTINGENCY

If scientific talent and resources were everywhere the same, the rate and character of scientific discovery might be more uniform. The high steppes of central Asia might have been the site of the first effective germ theory of disease. But it wasn't. In 1800, the electric-chemical discoveries made by Berzelius in Stockholm might have been made by people living in Constantinople, Tokyo, or any of the largest cities in Africa. But they weren't. The pattern is decidedly nonrandom. All possible discoveries are not equipossible. The aspects of a scientist's conceptual world that produce departures from equipossible discovery may concern history, culture, climate, financial sponsorship—whatever facts support ingenuity, sustain hard work, or impose disciplined training. But, importantly, none of these features get traction unless the discovery promotes a theory or theoretical outlook that actually captures the causal structure of the world.

History is not an experimental science, and so we cannot nail down that a scientific success is genuinely contingent rather than, say, inevitable because redundantly determined. But there are signposts when an event that can be tied to a theoretical advance occurs against a background of a normal or routine course of events. For example, further examples of small contingencies, of unplanned advance, can arise from factors beyond the internal sources of scientific investigation. The brewery next door prompted Joseph Priestley to mull over apparently "heavier air" emanating from that location, and that became an occasion to study the features of carbon dioxide. Priestley was especially interested in its ability to extinguish glowing embers, while at the same time he uncovered the powerful combustive forces of oxygen largely by chance. As he characterized the contingency: "If, however, I had not happened, for some other purpose, to have had a lighted candle before me, I should probably never have made the trial."[19]

In fact, Priestley offers a lovely general description of contingency:

> The contents of this section will furnish striking illustration of the truth of a remark which I have more than once made in my philosophical writing and which can hardly be too often repeated, as it tends greatly to encourage philosophical investigations; viz. that more is owing to what we call

> chance, that is, philosophically speaking, to the observation of events arising from unknown causes, than to any proper design, or preconceived theory in this business.
>
> For my own part, I will frankly acknowledge, that, at the commencement of the experiments recited in this section, I was so far from having formed any hypothesis that led to the discoveries I made in pursuing them, that they would have appeared very improbable to me had I been told of them; and when the decisive facts did at length obtrude themselves upon my notice, it was very slowly, and with great hesitation, that I yielded to the evidence of my senses.[20]

Some contingencies vault an entire field forward. Of course, this all didn't just "happen." Lucky breaks must be coupled with minds able to realize their importance. Pasteur was deeply engaged in the science of his time, and when you are busily employed in the business of inquiry, you are in a position to benefit from fortuitous opportunities. Pasteur modestly, but probably accurately, put his discoveries down to this benefit. In his words: "Dans les champs de l'observaton, le hazard ne favorise que les esprits préparés." ("In the field of observation, chance favors only the prepared mind."[21]) This is a common theme that acknowledges contingency. Joseph Henry, the distinguished American physicist, said, "The seeds of great discoveries are constantly floating around us, but they only take root in minds well-prepared to receive them."[22]

TYPES OF CONTINGENCY

The success of modern science doesn't issue from only the most deliberate methods of investigation; if that were true, science would have taken off when we developed mathematics and had the resources to perform voluminous observation. But it didn't. Instead, our explanations succeeded because our background theories were very good, when we made the right leaps. And we can arrive at those theories for many reasons, not all of them "rational." Thus. the lessons of contingency support the version of Inference to the Best Explanation defended earlier.

There are at least six kinds of contingency familiar in human history: psychological, environmental, timing, historical, cultural, and biological.

1. *Psychological Contingencies (Idiosyncrasy).* Scientific advances sometimes originate from personal variables—if not outright idiosyncrasies. And this fact can have confirmatory value. As noted earlier, it is a canon of experimentation that evidence is more secure when established by diverse means. Idiosyncrasy is one source of diversity in testing, a source of unplanned variation. Thus, idiosyncratic concepts and practices serve the purposes of intellectual foraging. The influence of idiosyncrasy should not be surprising, because little is known about the discovery process. What do such individual accidents look like? Philosopher of mind Jerry Fodor captures perfectly the experience of unexpected insight:

> The ways that people do this are notoriously idiosyncratic. Some go for walks. Some line up their pencils and stare into the middle distance. Some go to bed. Coleridge and De Quincy smoked opium. Hardy went to cricket matches. Balzac put his nightgown on. Proust sat himself in a cork-lined room and contemplated antique hats. Heaven knows what De Sade did.[23]

People clearly do things they believe will put them in creative moods. The techniques are as varied as personality. Indeed, there are many subjective reports of moments of insight in the history of science that have no detectably objective or rational basis. Instead, it is largely based on a person's quirks—his or her peculiar sleeping habits, social incompetence, unusual training, and so forth. Kuhn, for example, claims that Kepler was a sun worshipper, and this eased his transition to a sun-centered Copernicanism.[24] As in this case, these idiosyncrasies translate into success only when they reliably connect to theoretical features of the world, and conversely, personal quirks may just as easily cause people to *miss* opportunities to develop accurate theoretical views. The important point is that success, like failure, can have a chancy and personal beginning, and it is no argument against a given theoretical view that it began with either a dutiful experiment or a hallucination.

2. *Environmental Contingencies.* So much of the history of science is told as a character narrative, a study in how intellect bent the world toward progress. But just as often, scientific progress occurred because the world was already bent. Environmental contingencies are geographic features of the world that made science and its discoveries possible or likely. Scientific advance awaited leisure and concentration of resources, and this required civilizations. But the rise of dense civilizations normally depended on not one but many factors, not all of which are environmental. The presence of local water may be one of many properties that contributed to a thriving civilization, even if the civilization wouldn't have failed without it. And its presence may be an environmental contingency. The proximity of water is often thought

to be a necessary condition for a civilization's prosperity, but the abundance of water is only an occasional presence in rising or flourishing civilizations. Many great civilizations, like Athens and Constantinople, were not near substantial bodies of fresh water. Sometimes access to abundant water contributes to a civilization's success as much for trade as for agriculture and drinking. In this way, scientific development depended on distinctly geographical, practical contingencies. This is not to say that all scientific centers were near significant waterways. Some civilizations near water haven't made scientific advances, and some cultures not near significant water sources, like Madrid, had flourishing science. It is to say that the presence of water typically plays a potent causal role in producing the growth of scientific culture. Nor is abundant water sufficient for a flourishing civilization. Witness that the largest US cities are not near the largest US rivers (measured by discharge), and relatively few, like Chicago, are on a lake. The ultimate success of a large, civilized center can depend on a vast variety of factors—geographic, cultural, economic, and theoretical. This is precisely the multicausal path that contingent processes follow.

3. *Contingencies of Timing.* Of course, not all guesses, accidents of history, and chance meetings result in a theoretical advance. More often than not, they led to false belief and missions down rabbit holes. And it is the causal structure of the world that determines which cases of scientific serendipity result in findings that are durable and unifying. In fact, we assign more confirmatory weight when a serendipitous or unexpected event connects unrelated theories. Some of them concern uncanny timing. Usually these lucky events are treated as mere intellectual curiosities, sources of

amusement. We benefit from these fortunate happenings, but precisely because these events are so chancy and unexpected, no one can hope to *use* (or even rely on) fluky timing as a reliable pivot point to produce discoveries. But when the effects of contingent timing are beneficial, it is an especially powerful form of confirmation. Here is a typical case of fortunate timing, when in 1846, Claude Bernard

> solved the mystery of the carnivorous rabbits. Puzzled one day by the chance observation that some rabbits were passing clear—not cloudy—urine, just like meat-eating animals, he inferred that they had not been fed and were subsisting on their own tissues. He confirmed his hypothesis by feeding meat to the famished animals. An autopsy of the rabbits yielded an important discovery concerning the role of the pancreas in digestion: the secretions of the pancreas broke down fat molecules into fatty acids and glycerin. Bernard then showed that the principal processes of digestion take place in the small intestine, not in the stomach as was previously believed.[25]

Had Bernard not noticed the clear rabbit urine, he never would have asked the question of why they were passing clear urine. Had he not known about carnivore urine, he would have never pursued his explanation. Had the rabbits not been denied food, he would not have asked whether digestion occurs beyond the stomach. And importantly, if Bernard had seen the rabbit urine *before* having the background knowledge, the discovery may not have occurred.

4. *Sweeping Effects of Simple Historical Contingencies.* Some tiny events have unexpectedly large consequences when, for independent reasons, geographic or political

conditions align. Or so it has been claimed. Lynn White spelled out the important, yet contingent, consequences of the stirrup for Europe.[26] There, the introduction of the simple stirrup changed military history, and thus the history of a continent. Soldiers on horseback no longer fell from missing a blow, and so could swing their swords more frequently and with greater abandon. This new-found stability on horseback gave them a decided advantage over their less wieldy opponents. No matter what one thinks about this particular case, there are large historical impacts from local events, like the Venetian Navy's decision to pursue the use of telescopes for early detection of enemy ships. This may at first appear to be an insignificant decision, but it prompted Galileo's refinement of lenses, and his use of them together in more powerful arrangements in telescopes.[27] At that point, they were easily and fruitfully turned to the sky, for celestial observation.

5. *Unintended Cultural Side Effects.* Culture is dynamic, and its developments can throw off powerful side effects, like pollution or processed foods. An example of unintended cultural side effects comes from the Islamic world, in which timekeeping was crucial to prayer. Not surprisingly, resources, attention, and ingenuity were applied to the challenge of accurate timekeeping. But it was, of course, a contingent fact that a religion arose that placed such a priority on accurate timekeeping.[28] That is, timekeeping was not developed for its own sake, no matter how useful it may actually be. Only in service of a religion was such an important advancement achieved. Another example is the Industrial Revolution and the rise of thermodynamics—the steam and the internal combustion engine, as well as electric

generators, all were developed to assist in the push toward industrialization, rather than for their own sake.

6. *Biological Contingency and Cognitive Limits.* Another kind of contingency is biological, and it comes from the distinctive limits of a species' brain—its computational power. These computational limits are analogous to the Umwelt discussed earlier in the chapter: while there is a particular number corresponding to the insects a spider catches in its web every day, the spider will never know it; spiders have no concept of number. Yet it is a fact that there is a concrete number, and that number has causal consequences. This number is a part of an explanation of the spider's continued health—its adequate nutrition—that is a true explanation that the spider will never know. Just because it's correct doesn't mean the spider has access to it. Like the spider, we too may be ignorant of some facts about ourselves—like the causes of human consciousness—that we may never be able to appreciate. There are also experiences had by other species that are closed off to us. While we know what it is like to feel jealous, to get angry, or to experience joy, we haven't a clue about the feeling of the lungfish's first breaths out of water, or the feeling that draws a lemming into group behavior. When an organism's experience is so remote from our own experience that we can't understand what it is like to be that thing, we are perceptually or cognitively sealed off from it. While we are aware we don't share other species' Umwelts, that knowledge alone gets us no closer to understanding their experiences. Given that there are things we *know* we don't know (e.g., how consciousness works) and that there are things we know we will *never* know (e.g., other species' Umwelts), it is clear that humans

can't be the pinnacle of *possible* intelligence, understanding, or computational power.

In terms of advances in our scientific understanding, it is a contingent fact that humans have rigid neurological limits on how much we can remember, how quickly we process information, how much information we can process at once, and the complexity of ideas that we are able to entertain. The contingency of cognitive limitations is important because it suggests strongly that there are true explanations that we can't understand. But I don't have to defend this strong claim. It is enough to argue that there are explanations that we don't understand at the moment. If we characterize explanation as a description of underlying factors that bring about an effect, then it is possible to have explanation without understanding, because it is possible to accurately describe causal factors without a receptive audience, or any audience at all. In fact, this ontic conception of explanation, the idea that there can be explanations that are true even though no one understands them, is marked by its insensitivity to audience.

I want to develop this ontic idea that an explanation's goodness is dependent on its accuracy or truth, not on whether anyone does or will understand it. The ontic view is not about the nature of *explanation*; it is a view about what makes an explanation *good*.[29] In keeping with this ontic view, there are perfectly good explanations that may *never* be understood. Certainly there are some that *aren't* understood. And this cluster of issues raises questions about the metaphysical status of such explanations. Are they abstract objects? Creatures in Plato's heaven?

Here is the expected metaphysical question: If an explanation has no audience, then in what sense is it an explanation? If there are explanations that no one ever thinks or even proposes, what are they? In trying to account for what makes an explanation good, the objective, ontic view treats the assertion that an explanation can be good without anyone around to understand it as a superficial but still useful way of characterizing explanation. On this view, explanations could be any number of things: actual or possible taxonomic arrangements of relevant causal details, for example. And if language is suitably naturalized, then the fact that these possible or actual causal arrangements are expressed by linguistic items that have pragmatic, semantic, or syntactic features is no obstacle to inclusion in the physicalist's inventory. So there is no need for Plato's heaven. Calling explanations true even when lacking an audience is a manner of speaking. Saying that you can have a good explanation that no one can understand is a bit like saying, to borrow a saying from Quine, that an ancient tide is clockwise. And depending on the reason for absence of an audience, other images might be more appropriate. Some uncognized explanations might await a proper arrangement of causal details; they are unflipped coins of sorts. Other explanations could be uncognized because they never became accessible and are now lost forever, in the same way that there might be evolutionary truths whose only evidence is in soft tissue now long perished and so permanently missing from the evolutionary record. There are many acceptable options here, and the account of ontic explanation I offer in no way depends on which we accept or, at this point, *that* we accept one.

Experts may reach explanatory limits because there are too many mechanisms to track, while for laypeople it might be that *and* that the detailed ways in which even a few mechanisms interact are too arcane to encode. These cognitive limits create one class of reasons to idealize and simplify. When experts model biological systems, they make all sorts of simplifying assumptions about whether the population is panmictic, or whether a gene is Mendelian, or whether the organism's community is Lotka-Volterra. Departures from these ideal assumptions are difficult for a normal human mind to trace, both because there are a large number of components to attend to and hold in memory and because their effects are not independent; they are not straightforwardly additive. For laypeople, it is possible to memorize for an exam, say, the active ingredients of an antibiotic, and yet not grasp the chemistry of bactericidal drugs. In this kind of common case, there is a truthful explanation of how antibiotics frustrate infection, but not one that this student understands, even if the student knows the things an explainer often knows.

Many people, however, would *define* explanation in terms of understanding. To them, a good explanation succeeds at increasing our understanding of some event, or at least aims at doing so. In particular, understanding is said to be characterized by the ability to describe the causal features of an object. That kind of effort takes a familiar form. Understanding an event usually involves understanding its parts. Achieving a better understanding of a thunderstorm demands that we understand what a cumulonimbus is, the charges in different parts of the cloud, the air currents within the cloud, and other features that account for its behavior. In short, we achieve an understanding of an explanation by attaining ever better, more accurate descriptions of the mechanisms and processes that bring things about.

It may be that there is no understanding without explanation.[30] But it is also true that there is also no understanding without memory, without attention, and without awareness. That's why we deny protozoa the capacity for understanding but credit primates in some measure with it. To understand a problem, for example, we must be able to focus on some part of the problem and relate it to other parts. Sustained focus requires that we remember the details of those components for as long as we need to in order to represent the problem.

If this is the case, then we should ask what the limits are on the human capacities for understanding, which means asking what the limits are on memory, attention, and awareness. With all these components, it is not surprising that understanding is a graded notion; it is not an all-or-nothing affair. So what would be required to say that we "understand" any given distinctive feature of the world? Minimally, we must be capable of representing the many components of things like, say, poverty or antibiotic resistance, and that means suspending them in working memory. Very crudely and basically, that process invokes the activated portion of long-term memory, the focus of attention, and a set of memory buffers that hold information for very brief time intervals. The need for speedy processing imposes forbidding constraints on memory and attention, the two chief faculties underlying understanding.[31] But there are also limits on the number of items in memory that can be concurrently activated,[32] the number of relations among items in memory that can be integrated,[33] and the limit enforced to avoid interference between items in working memory.[34] We can barely add the cost of each grocery item as it passes on the conveyer, let alone retain the cost of each and weigh them in order of causal contribution to the effect we want to explain. Why should we expect to be any more successful if these are also the requirements of understanding? Unfortunately, we shouldn't: the fleeting temporal windows of memory and attention, combined with the vast number of component processes in the systems we want to understand, create a processing load and a scale that cannot be tracked by introspection.

Developing this naturalistic account is important because while we occasionally surmount some of our biological limitations, we doubtless have myriad, perhaps infinite, natural biological limitations and imperfections. These biological constraints entail that some objective understanding is incomplete, and so we need an account of incomplete understanding. It is likely that a sense of understanding influences the acceptance of an explanation: one might reject an explanation due to a low feeling of understanding or accept one due to a high feeling. Unfortunately, as we have seen, a sense of understanding doesn't necessarily track a true explanation. Reber and Unkelbach documented a similar point for beliefs about the truth of assertions (or beliefs).[35] Exposure to a statement like "Lithium is the lightest metal" increases its perceived truth by increasing its processing fluency. That is, the mere repetition of an item creates a familiarity that promotes a high feeling of understanding. Yet this goes not just for statements that are false, but also for statements that subjects have been *told* are false. Recall the discussion of a fluency account of explanation. The repetition that produces processing fluency continues to have its effect, whether or not that statement is true.[36] And so flourishes the sense of understanding, independent of truth.

An entrenched theory creates precisely this kind of familiarity or fluency, and with it, perceived truth or sense of understanding. But across great causal distances a sense of understanding might be difficult to sustain. For example, some explanations are complex because the mechanisms they describe are spread out in space or time, or located in a contorted network of causes: they are "long distance" explanations. Reasoning about cross-domain causal relations is hard even for experts.[37] However, many important domains of scientific study are characterized by cross-domain ("long psychological distance") connections, such as genetics and mental diseases, human activity and global warming, socioeconomics and obesity. What factors contribute to sense of understanding and Acceptability of Explanation (AoE) for such explanations?

Does exposure to multiple "long distance" causal explanations increase sense of understanding and AoE of a subsequent "long distance" explanation? For example, even if one would normally have a low sense of understanding and AoE of how the vibrations in air get turned into the perception of noise, might sense of understanding and AoE increase after being exposed to explanations involving other mind-brain connections? For the target explanation of hearing, would only other sensory phenomena (e.g., taste and sight) produce a transfer, or might more distant phenomena that still cross physical-mental domains produce transfer (e.g., genetics and addiction, or environmental chemicals as triggers for mental diseases)?

To maintain some sense of cognitive control over taxing explanations, we embrace unduly simplistic models to explain otherwise mysterious effects. Traveling by long-distance explanation, we find positive evidence that urinary incontinence can be controlled by biofeedback,[38] that birds "see" electromagnetic waves, and that the dung beetle navigates by stars.[39] Not all of these long-distance effects are healthy. Some whales beach themselves in reaction to low and midfrequency signals miles away,[40] and for humans, the chemicals in smog appear to have temporal as well as geographic long-distance effects on health, among them, premature death.[41]

Causal distance is what separates a cause (or set of causes) and the effect we want to explain. Causal distance imposes a barrier to understanding when very different kinds of causes seem to combine to bring about the effect. What is the connection between Crohn's disease and cleanliness (as the "hygiene hypothesis" suggests), the velocity of light and Brownian motion, or the relational versus nonrelational explanations you prefer and the kind of crops you grow?[42] When complex natural systems include numerous mechanisms from otherwise disparate domains, it is more difficult to appreciate their functions and synthesize an explanation. It is one thing to understand that pressure increases when a gas is heated in an enclosed container. After all, the components are small in number, and only a few of their properties (in this case, of molecules) are explanatorily relevant. Contrast this with explanations of the behavior of complex systems like conscious processing, crime, educational achievement, and watershed pollution (to name just a few) with tens or hundreds of factors whose performance cannot be tracked anecdotally, but only with elaborate computational models.

This investigation is designed to be consistent with a scientifically credible theory of understanding that will explain what must go right when we actually understand, as opposed to merely feeling that we understand (perhaps incorrectly). Having a theory of understanding would be a significant accomplishment. But combined with the idea of contingency, this account should also explain theoretical progress in science, a stubborn obstacle in formulating explanations for the history of intellectual achievement. With a more graded, empirically responsible account of understanding, we can ask how mistaken conceptions can lead to scientific progress. How is it, as I claim, that the rise of modern science is owed to a Newtonian hunch about the ballistic composition of matter, inherited from a dying, late 16th-century corpuscular alchemy that was surely incorrect?

Initially, this claim seems exotic, but it could be made less dramatic with a fuller theory of understanding. Very schematically, such a theory would trace the way in which alchemical relations were represented in the mind by the mechanisms and processes of the brain's executive function: attentional windows in the tens of milliseconds, an attentional buffer in which information can be held, the rate at which the acuity of that information degrades when suspended in that buffer, the number of items processed, the number of discriminable features of those items, their respective modal sources, and their role in promoting or inhibiting an effect. When these components' representational features covary in the right ways with states of the world, we can more fully *understand* true explanations. When you understand the causal contributions of each causal factor in an effect, we are able to identify and track the difference that each factor makes to it. The sensitivity of this cognitive system to changes accounts for the importance of "difference making" features in some accounts of explanation.[43]

In the end, our Umwelt and our natural cognitive boundaries shape the sorts of explanation we can understand. Between these two restrictions on understanding, it is possible, even appropriate, to accept an explanation that is true but, due to barriers of our Umwelt or our cognition, does not convey understanding; it cannot be fully understood by us.

To illustrate the gap between explanation and understanding, consider the phenomenon of inheriting genetic traits. The source of heritable traits was long a mystery, well before Darwin, and even before Mendel. But it remained uncertain into the 20th century, when the great biologist J.S. Haldane voiced his deep reservations about the idea that our hereditary characteristics could be carried by some "germ-plasm" or physical substance in the nuclei of cells.

> The real difficulty for the mechanistic theory is that we are forced, on the one hand, to postulate that the germ-plasm is a mechanism of enormous complexity and definiteness, and, on the other, that this mechanism, in spite of its absolute definiteness and complexity, can divide and combine with other similar mechanism, and can do so to an absolutely indefinite extent without alteration of its structure. On the one hand we have to postulate absolute definiteness of structure, and on the other absolute indefiniteness.
>
> There is no need to push the analysis further. The mechanistic theory of heredity is not merely unproven, it is impossible. It involves such absurdities that no intelligent person who has thoroughly realized its meaning and implications can continue to hold it.[44]

Haldane's conclusion is understandable. He found the mechanistic theory inconceivable simply because he didn't have the theoretical resources to conceive it. For the mechanism of heritability to be physical, it must be able to achieve two feats: (1) regulate the development of extremely subtle and complicated biological systems and (2) retain its power to regulate the virtually endless divisions and combinations through successive generations of organisms. He knew of no such ordinary physical substance that had these properties, and so concluded it was impossible. Thus, prior to uncovering the explanation for genetic inheritance, in the second decade of the 20th century, Haldane found himself lamenting our lack of understanding of heritability: "It seems perfectly clear that germ-plasm of so simple a character as this could by itself furnish no explanation whatever of the development from it of the adult organism with all its enormous complication and absolute definiteness of structure."[45]

And yet, this explanation is precisely what the germ-plasm began to furnish. Just 40 years later, the mind recoiled from the opposite assumption. By 1953, Watson and Crick had discovered the double-helix structure of DNA composed of two strands of simple nucleic acid "bases" with the crucial feature that there is a unique correspondence between the bases on the adjacent strands. Both DNA and the related RNA molecule were the chemical substances that carried genetic information and that the particular sequence of bases in a strand of DNA "coded" the required genetic information.

The ontic account of explanation itself accounts for the precise character of progress in the history of science. Sometimes accurate explanations were widely rejected because people didn't have the conceptual or theoretical resources to appreciate its truth, as happened when Haldane rejected a genetic account of heritability because he couldn't conceive that the mechanism of heritability was physical.[46] In fairness, Haldane appreciated the limits on human cognition: "My own suspicion is that the Universe is not only queerer than we suppose, but queerer than we can suppose."[47] So Haldane was acutely aware of our contingent, cognitive limits. And sometimes theoretical progress was made when an accurate explanation was accepted and pursued even though its features were only dimly appreciated at the time, as was the case when Watson and Crick discovered the double-helix structure of DNA.

These scenarios of science carry at least one lesson: scientific progress is a routinely contingent process. In ordinary scientific practice, scientists project and pursue hypotheses with incomplete information—that is the point of the hypothesis to begin with. The extent of the ignorance varies, but when progress occurs, vaulting that gap of ignorance is normally a result of contingencies of scientists' psychology, funding, timing, geography, and epoch. In most of these cases, experimental knowledge does not serve to make this gap compact. The fact that a contingent discovery could move science forward is a testament to the traction of an unconceptualized, unobservable, mind-independent reality. The fact that we are ever forced to abandon a theory we constructed, in the face of evidence, is a sign of this reality's force. The first hint that there is a resistant world behind the appearances—a world that pushes back—is the evidence of the world before minds, and thus a world that minds may not understand.

THE COGNITIVE BASIS OF EPISTEMIC BOUNDARIES

We are not surprised that there are true explanations that the tick, fly, shark, bird, and monkey don't understand. Why shouldn't we think that there are true explanations that we can't comprehend? After all, as occupants of the phylogenetic continuum, we are little different from our closest neighbors, and there are true explanations of their behavior that they can't comprehend. With company as distinguished as dolphins, monkeys, pigs, and chimps, why would we think that humans, alone among creatures, face only explanations we can understand, that there is nothing beyond our mental reach?

Consider three cases of explanations potentially shaped by our cognitive limits:

1. *Computer simulations of multivariable systems.* If you have a model of the ecology of a polluted watershed (normally implemented on a computer), there are tens or hundreds of variables in the model. They might include facts about dozens of toxic chemicals from agricultural runoff, road surface runoff, night-running chemicals, poor drainage, high water table, industrial waste, algae blooms, flow levels, the effects of channelizing, and so forth. Each of these causes may only contribute a little to the overall effect. Some of these effects may be interactions, too. So the question "What explains the pollution of this watershed?" may invoke many causes, too many for human understanding to fully capture in a gaze, and too many to be assimilated to any simple model of understanding we use in ordinary cases. The fact that we possess written and electronic records doesn't eliminate the need for understanding but dramatizes the distance between inexpert and theoretically trained understanding. It also highlights how little we understand when looking at a complex numerical output, even though we have the concept of number. As Nancy Cartwright observed,

there is sometimes an inverse relationship between truth and usefulness.[48]

2. *Consciousness*. Humans have inbuilt cognitive limitations that place some explanations permanently beyond their grasp. This is especially so for accounts of explanation that demand much from understanding. Our faculties of attention and memory arise from a rigid neurological architecture. The ordinary standards of understanding—normally requiring that the causes responsible for the effect be introspectable and separately trackable—cannot be met by an architecture that squeezes efficiency out of processing by rendering so much processing automatic, shallow, and opaque. Our perceptual and cognitive systems are capable of maintaining only very small numbers of memory and attention contents from just a few information sources. If we had undistorted and deliberate access to the content of attentional windows of arbitrary size, could track those events with less constrained powers of identification and discrimination, and could integrate the many temporal levels and modal qualities of information, we might be able to meet the standards of grasping or understanding normally imposed when we attempt to "understand" conscious experience. In other words, if we had capacities that we don't, we might have been able to do things we can't—in this case, understand consciousness.[49]

This grim prognosis for understanding consciousness is not unique to consciousness. This situation is routine in other sciences that model multivariable systems. Consider prognostic risk models for chronic health conditions like colon cancer. These models can potentially have hundreds of candidate predictor variables, including demographic factors (age, sex, race, education level, region of the country), genetic profile factors (sometimes dozens of them), and behavioral factors (nutrition, exposure to certain chemicals, etc.). For many diseases, most of the variables are weakly predictive, but no single variable is a good predictor on its own. So the question of what explains the occurrence of colon cancer is normally answered by appeal to a cognitively untrackable collection of variables documented by a printout of compiled or opaque information rather than a recitation of introspectable states that we can attend to separately and trace independently and jointly to their effect. And in fact, we are so ill-suited to this cognitive task that it's pretty routine to use some kind of statistical variable selection process to prune the number of variables in the model to arrive at something that's "clinically interpretable"—in other words, to work around the limits of our processing ability! No doctor would be able to quickly use and understand a model with 200 variables in it, even though that model is probably more predictive than a model with only 20 variables.

The neuroscience of consciousness, too, will produce a complex computational theory that similarly outstrips our unaided cognitive powers. This analogy of explaining consciousness and explaining colon cancer is not designed to persuade researchers in the field, but to enumerate the natural cognitive limitations that make our efforts to understand and explain consciousness and colon cancer so frustrating. These limitations explain our greatest dissatisfactions with physicalist accounts of consciousness. A common objection to physicalism, the idea that there are no things other than physical things, is that it could never explain how subjective, conscious experience could arise from something purely physical. But many philosophers have noted the peculiarity of claiming that there is something wanting in the physicalist explanation for consciousness. Even on a complete list of items in the universe, we can't imagine at some point during the reading of this list that we say, "Oh, that is how consciousness is created!" or even, "Oh, that explains why spicy foods feel that way!" (even if we know what makes something spicy, that's a different question than why it *feels* the way it does). People often say that physical items are just not the right kind of thing to produce subjective experience, but the fact is that they are saying it without having even the first idea how many causal relations and items are involved in ordinary awareness. So in effect, they have a very specific conception of what an acceptable explanation would look like. But what, exactly, are they imagining would make them find any explanation credible? This is the first instance of two kinds of cases in which people's *conception* of explanation may be the obstacle to accepting a good explanation.

3. *Quantum effects.* Sometimes when physicists are looking at quantum effects, the best explanation is that the phenomenon is irreducibly statistical. So they are looking at a probability distribution, sometimes many pages long. Now, if all there is to a quantum effect is this ensemble of probabilities, then this ensemble is a kind of compositional explanation of the effect. It is not a causal explanation, but one that explains what makes up or constitutes the phenomenon. It is also the only type of explanation that exists in this case. If you cling to a different account of explanation, then you might reject that long list of probabilities as an explanation, expecting something more. Or take the experiments establishing quantum uncertainty. In them, a particle is either spin up or spin down. There is no explanation

available to us for why the particle is spin up, and no explanation available to us for why it is spin down. In fact, it is one of the grand insights of 20th-century physics that it might be folly to try to explain it. If your conception of explanation finds this lacking, perhaps it's your conception that's wrong.

These episodes are nearly impossible to explain unless we suppose that the world regularly pushes back against our beliefs. Empiricist philosophers of science argue that we can only use information about observable phenomena to correct a theory. Social constructivists like Kuhn—who include theoretical statements on their list of socially constructed knowledge—claim that scientists who work in different paradigms live in different worlds, and thus, for example, the worldviews of Newton and Einstein were incommensurable. He asserted that "after discovering oxygen Lavoisier worked in a different world,"[50] that "after Copernicus, astronomers lived in a different world,"[51] and that "until that scholastic paradigm was invented, there were no pendulums, but only swinging stones, for the scientists to see. Pendulums were brought into existence by something very like a paradigm-induced gestalt switch."[52]

Of course, after these discoveries, scientists live in different psychological worlds, and that brings with it effects on the physical world. Minimally, they have different beliefs about the world, and that constitutes a psychological difference. But it does not mean that scientists who work in different paradigms live in different mind-independent, physical worlds. If paradigms had that kind of world-constituting power, and the conceptual background of relativistic kinematics caused Einstein and Newton to "live in different worlds," then relativity theory would be denied the confirmation generated by Newtonian observations of conserved mass approximating the effects of convertible mass at low velocities. The same pattern of confirmation applies to observables as well. For example, most of the comet sightings from 239 BCE to 1222 CE were made under the aegis of the ancient Greek theory or the ancient Chinese theory, which is how the orbit was determined and fixed. If observation were actually constituted by theory—and so a modern theory could not rationally recruit objects observed under a historic, discredited theory—then we would have to deny the hypothesis of the orbit period an important source of confirmation.[53]

The accuracy of many of these epistemically progressive, successful theories is fortuitous. If you can shoot someone without knowing who it is, and you can refer to someone without knowing who they are, you can surely explain something without knowing what you are explaining. This may sound funny because the value of explaining isn't plain when your audience doesn't understand what you are saying. But supposing that explanation is successful only when the audience understands makes it difficult for bounded minds to embrace complicated phenomena that go beyond our contingent cranial limits.

Empiricist positions that focus on piecemeal accumulation of observations have vastly underestimated the importance of contingencies in the history of science: the psychological, historical, temporal, biological, environmental, and cultural. Contingencies large and small tear us from our comfortable prejudices and move us in a fortuitous direction. True explanations that we don't now completely understand remind us that there is a reality beyond our brains. There is something true beyond what we can understand now. Contingency, explanation, and intellectual progress are all closely aligned. Once we dramatize the contours of the human mind as a source of contingency, we have to accept the consequences of this contingency for explanation as well.

There are explanatory gaps between physical descriptions and psychological ones, between older theories and newer ones, and between beliefs and the unconceptualized realities they describe. Each gap may present an intriguing puzzle, but they all arise from the same challenge: to understand the world beyond our brains. But humans are not the final link in the great chain of being. Just as surely as there are facts that a frog can't know, there are facts that answer mysteries looming beyond the uneven contours of human boundaries. Individual differences in personality and intelligence may shape cognitive boundaries. For example, a normal person may still be less empathic than others and fail to understand fully why others seem so wounded by frank negative appraisals of their performance. And, regardless of our varying empathic limits, there are some facts that none of us could understand, even about other human beings whose cognitive makeup may be very similar to our own. We couldn't possibly appreciate, except in a primitive way, what a day in the life of a medieval peasant or 19th-century Trobriand is like. There are true answers to these questions about their lives, and people explaining their behavior would invoke these truths. These sorts of cases are perfectly ordinary, and it is not in the least difficult to see why they illustrate routine and ubiquitous limitations on understanding.

It is embedded in my view, then, that we can explain things that we don't fully understand. We can explain electron spin in a way that will correctly orient research for decades, and explain it beyond our conceptual resources to fully represent it. We can explain something by describing causes, even if doing so only advances understanding in tiny increments or isolates the causal roles of objects whose characteristics we can only partially articulate. Late 16th-century alchemy began isolating the causal role of "corpuscles" that were, in fact, the elements of modern chemistry. Late 19th-century ether theory began to isolate what was in fact the electromagnetic field. Early 20th-century physics began to track that portion of mass that is conserved and the portion that is convertible with energy. Scientists of the time may not have conceived of their objects in this way. But our task for today is to explain not scientists' conceptions or their "mental models," but instead their achievements. Contingent or accidental events can help us to traverse explanatory gaps. But we can explain this only if the world is the kind of thing that can push back.

CONCLUSION

Our cognitive limits may make true explanations intellectually unsatisfying, offering little more than a bare appeal to an unobserved world. But the world sometimes compensates for that feeble appeal, because it allows us to build out from that feeble image. When Peirce likened understanding to a key turning in a lock, he was assuming that the mind's teeth could engage the world's tumblers. But this image isn't everywhere true. There is a world at 30,000 Hz that is silent to humans, but alive to bats and crickets. We know that world not through hearing, but through other, more indirect modalities. We know it through cellular recordings from the bat auditory cortex and from cricket communication. Regional deformation happens slowly, behind our backs, as the epochs race by. We know it through radioisotope dating. The same goes for evolution. Without a fossil record, the evidence of evolution would be even more inaccessible.

If we are to adequately explain the success of science, its historically asymmetrical ascent punctuated by leaps, explanations had better be evaluable by standards that transcend culture and epoch. That is, we will want to be able to say that Paracelsus had a poor explanation for circulation, and William Harvey had a better, more accurate one, full stop. It might be tempting to say that the Paracelsean explanation for circulation is a good one *for them*, but not for us. Yet to do this, we lose the ability to speak in terms of progress.

When we gaze across the history of science, the terrain seems stable enough to allow detection and description by earnest investigators, to identify fraud and genuine discovery alike. What are the chances that a moon landing could be faked, or a radiofrequency ablation treatment could end arrhythmias? The first contingency is so easy to refute, and the second to confirm, in a world that pushes back hard against the theories we advance. It is a very good reason for supposing that we have correctly identified the cause of an errant electrical signal propagation in the heart given that when you burn the tissue carrying it near the source, the arrhythmia stops. That fact about the world is stable; it doesn't change to suit humans. What changes is humanity's ability to understand the explanation. And that change, we have seen, comes from contingency.

If that weren't enough, contingency, too, drove the Newtonian Revolution. The fact that so grand and unified a theory emerged from the dank basements of alchemists and adepts just one professional category removed from sorcerer or huckster provides all the evidence you need that a most illuminating inference can come from shady pedigree. People are responsible for pitching theories, sure enough; but the world helps judge them, by sorting without artifice or apology.

This statement may seem unforgiving of human frailty and normal limitation, but it certainly isn't unfamiliar. As we've seen, we think nothing of making the same claim about "lesser" species. Bees can't appreciate the true explanation for their nectar-locating abilities. Spiders can't explain the strength of their own silk, and they wouldn't understand the true explanation even if you took the time to explain it to them. The limitations are utterly analogous in the case of humans. Only the most extreme species exceptionalist would invite someone to conclude that humans are without such limitations. A good explanation is an accurate one, whether there is ever a moment when we understand it. But by following its causal deliverances, we may push science along the right track, even without a sense of understanding.

6

NEWTON'S HUNCH

HISTORY HOLDS MANY LESSONS ABOUT the power of contingency, the force of accidental or unexpected events. Sixty-five million years ago, a chance event changed the course of animal history on Earth. A meteor hit the earth, setting off a chain of events that killed off the dinosaurs. We can even heighten the sense of contingency by marveling at the low probability that a meteor would hit the earth, and how unlikely it is that large mammals ever would have evolved if it hadn't hit. We may not know the exact numbers, but we do know that the events were improbable and their effects large.

The effects of contingency are no different in the history of science than in paleontology. Consider the unusual historical position of Robert Boyle and Daniel Sennert.[1] Famous for "Boyle's Law" and his affiliation with Newton, Robert Boyle's place in history is well known. Sennert, on the other hand, was an obscure German physician, a contemporary of Boyle. Sennert was on the faculty at the University of Wittenburg and spanned the Aristotelian theory of forms and modern chemistry through a corpuscular alchemical theory of matter. Surrounded by alchemists engaged in primitive experimentation, they both had conceptual resources and corpuscular theories that worked well together with ballistic theories of the composition of matter, from gas to solutions. Sennert embraced an empirically troubled Aristotelian theory of substantial forms, and Boyle a primitive taxonomy that personifies inanimate causes. And yet their corollary corpuscular views were right in just the way necessary to make theoretical headway. What is the probability that a mystical view about unobservables would have arisen at a time and place when we had someone like Newton to apply mathematics to the dynamics of microparticles, or a nascent chemistry to exploit just those features of alchemical corpuscularism? How many Boyles or Sennerts existed in the history of science but were never favored by fortune?

BEST EXPLANATIONS FOR BEST SCIENCES

For curious minds, it is natural to pursue an explanation for the success of our best sciences. As I hope to have shown, I think the best explanation for the empirical success of contemporary mature science is the approximate truth of our scientific theories, or at least the intellectual reliability of scientific methodology *given* the high quality of our background theories. Some philosophers find this pattern of reasoning circular. But that would be to mistake abbreviation for circularity. And it is a compressed pattern of reasoning widely relied on in scientific and lay life. Without this cooperative abbreviation, we would need to tame and collect the diverse evidence for good, powerfully unifying hypotheses every time we want to appeal to them. In his work on Avogadro's number, Perrin recounts this pattern of inference in the argument for molecules: "Our wonder is aroused at the very remarkable agreement found between the values derived from the consideration of such widely different phenomena."[2] The molecular hypothesis ushered in by Avogadro explains everything from electrolysis to Brownian motion. With such precise agreement in values across such different phenomena, "the real existence of the molecule is given a probability bordering on certainty."[3]

In spite of my hypothesis for the success of science, in recent history the best explanation we have been given for the shift to modern science was the Grand Narrative, in which Newton exploded on the scene and put an end to the pseudo-science known as alchemy, clearing the way for classical physics and modern chemistry. However, the truth is rarely so straightforward. In the 1970s, historians of science deflated the Grand Narrative at least a bit, by highlighting Newton's own extensive alchemical practices. It appears that in Newton's worldview, alchemy offset the defects of classical and contemporary theories of atomism. Alchemy became a repository for all that was left unexplained by the mysterious processes that mattered to Newton: how atoms cohere to make objects, how the life force or activity of animals arises from them, and how transubstantiation occurs.[4] But the scholarship on Newton's alchemy did not go so far as to give alchemy credit for scientific progress. How could a nonscientific study result in such great progress?

That credit to alchemy, however, seems well deserved, and teaches us a lesson in contingency that should be applied to explanations for the steep increase in scientific advance around the beginning of the 17th century. It was, in the end, two doctrines—tested by alchemy—that moved the thinkers poised between two worlds and ended the lingering influence of ancient views: (1) the extreme minuteness of the particles that composed matter and (2) the retrievability of original constituents. The alchemist Sennert, and not modern scientists, demonstrated both those things. In 1619, he showed that silver dissolved in aqua fortis (nitric acid) could be made to reappear as a precipitant, thus demonstrating (2), and in 1636, he dramatized the truth of (1) by passing a solution through a fine filter paper before "retrieving" the silver as a precipitant.[5]

Laboratory practices common in corpuscular alchemy of the late 1500s and early 1600s, like titrations and filtrations, showed that corpuscles were regarded as hard, minute, and able to be combined. By the time Boyle arrived on the scene in the mid-17th century, 50 years after the prominence of these laboratory practices, agitation was added to the list of corpuscular powers. Though the corpuscular conception did not create a rush to document the microstructure of nature, it did happen to capture enough of the actual features of atoms and molecules, and did so accurately enough that it spawned a research program unprecedented in its depth and scope. It was this fact that meant that science could move forward: it tracked the real world better than any theory before it. And this development was historically contingent: while the theory was not fully worked out, the fact that it was right in important respects, as we now see in retrospect, largely explains why Newton's dynamical theory took hold so quickly and successfully when applied to objects of any size.

The Steep Ascent, then, in fact preceded by about 60 years the one traced in the Grand Narrative of the Scientific Revolution that high school students and educated adults learned. Around 1600, alchemy was dying and Newton had not yet been born. But in death alchemy was throwing off its most beautiful shoots, developing a corpuscular or atomistic view that provided the basis of vigorous empirical testing and growth of physical and chemical theories for the years, decades, and centuries to follow.

Historians of science might have thought that Newton's attachment to alchemy, with its roots in Aristotelian elemental theory, might have prevented him from embracing modern corpuscularism. But the experimental turn in late alchemy had, in fact, settled that question long before Newton. As Thomas Kuhn, the distinguished historian of science, summarized it:

> After about 1630 . . . most physical scientists assumed the universe was composed of microscopic corpuscles and that all natural phenomena could be explained in terms of corpuscular shape, size, motion, and interaction. That nest of commitments proved to be both metaphysical and methodological.[6]

It also proved to be largely accurate, and key to the stunning progress that ensued. What did later alchemists and early physical scientists find as compelling evidence for atomism in the later alchemical demonstrations? Again, the question arises: Why Newton? Why didn't an elaborate astronomy or advancements in theories of algebra spark a revolution in physics and chemistry?

Alchemy, Serendipity, and Inference to the Best Explanation

These questions can be boiled down to this: why did the science of selected, populated regions of 17th- and 18th-century

Europe come to dominate *the science* (or if you prefer, the understanding of nature) of every other part of the world?

This is a natural question because, for most of human history, Europe didn't dominate. The land and cultures we now identify with Europe's legacy lagged behind countless intellectual and practical feats across the globe. On the grass terraces of Europe, humans had barely assembled villages as Chinese astronomy had flourished for centuries, and as Persians engineered water sources called "quanats" in what is present-day Iraq. In the more recent past, at the close of the first millennium—during what is often called the Dark Ages of northern Europe—a more southern band of Islamic culture stretching from Cairo to Cordoba enjoyed a golden age of intellectual achievement, including the invention of algebra, instrumentally useful theories of light, an elaborate astronomy that supported wonderfully accurate calendar making, descriptive physiology, and practical medicine. Many formal, mathematical achievements arose, more or less independently, in India as well.

History seems a torrent of activity when our minds sweep across the timeline; we think it proceeded at the pace of our thoughts that trace it. Yet while humans were certainly occupied with intellectual projects for the last 10,000 years, our deliberate efforts to understand the natural world was not a rapid flurry but a sleepy, stumbling meander. There were long stretches of practical medicine that may not have left the population better off, descriptive accounts of the human body that were passed down in a lore but not in, say, a surgical tradition. Even the advances of the great Copernicus and Kepler were largely based on observations of macroscopic bodies, not theories of unobservable bodies and forces. And so, while it might seem that without treating Kepler as a turning point Newton's work would have been built upon a faulty foundation, there is reason to locate credit closer to Newton's environment and time. My focus is on Newton's *theoretical* rather than observational achievements. In this respect, Galileo might have been a more theoretically important predecessor than Kepler. Of course, origin is never about the single intellectual hunch; theoretical successes typically depend on many factors. Kepler's theory, after all, was mostly about observables—it was not chiefly theoretical, and so not a good candidate for having predated Newton in a success of equal importance. Galileo, on the other hand, was compromised not by unobservability but by contingency; he had insights about microscopic particles, but no Boyle (or those that followed immediately after him) to connect celestial mechanics to molecular physics and chemistry.

Other advances were drably descriptive, and not explanatory. In the 2nd century AD, Galen described capillaries and circulation in the animal corpses he dissected,[7] in the 8th century Bede recorded observations of the tides,[8] and in the early 13th century Grosseteste produced mathematical descriptions of light's refraction.[9] These processes of description and observation are instrumental, not richly theoretical. Even so, they supported empirical advances. Whether through serendipity, dogged observation, or the sheer robustness of the phenomenon, all of these successes were achieved without the benefit of prior theory, or at least theory that demonstrably tracks the world.

But then something interesting happened in the late 1500s and early 1600s: the rate of significant instrumental and theoretical discovery took off across Europe—in what is now England, France, Germany, Italy, the Netherlands, and Denmark. An impartial reader has to be struck by one summary finding so momentous it is available at a glance: theoretical science—mostly physics and chemistry—begins a meteoric rise just before the time of Boyle and Newton, between 1550 and 1600, during the later period of corpuscular alchemy (Figure 6.1).[10]

When you aggregate all of the events worthy of positive mention in everything from comprehensive history of science textbooks to encyclopedic science timelines, you get this steep curve. The aggregation can be almost as unselective as you like; it emerges on a variety of conventions you might adopt about what events to include or exclude: engineering achievements, observational descriptions, and theoretical developments, to name a few. They can be discoveries that occur without a theory about unobservable phenomena, mathematical models of light or planetary motion, engineering achievements that have no explanation for the strength of the materials used, and useful animal husbandry and plant fertilization successes under badly mistaken theories. No matter how you select your events, this curve remains robust.

One simple explanation for its steep rise is that Newton's vision was approximately true—though more probably, it was the underlying corpuscular view of later alchemy that was approximately true. In either case, the explanation for its success hinges on its approximate truth. What do we mean by "approximate truth"? As always, this is a rough label, implying different levels of accuracy for different aspirations. If you are sinking a glass electrode into a cell, approximate truth is measured in microns. If you are butchering a cow, the measure is in inches.

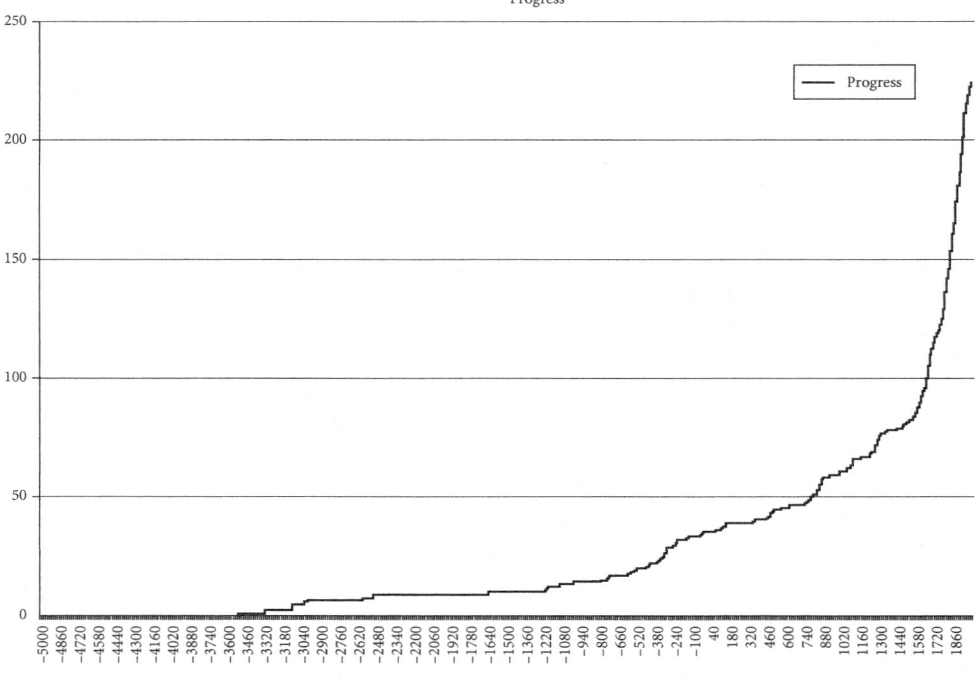

FIGURE 6.1 x axis = year; y axis = # of progressive events counted.

But this explanation for the steepness of the curve is too simple. After all, the curve could represent the benefits of network relations that increased in step with the population and social organization. With more people come more discoveries, and more people to share them with. So we might instead explain advances or progress in terms of the good effect of network relationships, such as professional organizations that allow easier cooperation and distribution of information across national borders. Such relationships across networks increase the probability that you will find someone who has an important piece of the solution to your theoretical problem.

What events created the network and sped propagation along its paths? The invention and commercial production of the moveable type printing press in the 1400s led to the rapid increase in the number and availability of core science texts. This made it possible for a greater number of students to study the core works, now that they did not have to wait to gain access to treasured manuscripts laboriously copied by scribes. At about the same time, education became more accessible, and the rolls of many of the great European universities swelled. With the rise of a public mail service in 1635 and the General Post Office in 1660, scientists could share results and correspond in an efficient way. Finally, just a decade into the second half of the 17th century, the scientific academies arose. All of these developments placed scholars and scientists in closer contact, increasing the probability that a scientist with a problem would find someone with a solution.

And yet, regardless of these technological and networking innovations, the scientific theories that accounted for the steep curve would not survive under diverse tests—at least not for very long—if they weren't based on accurate theories. In fact, the more contact scientists had with one another, the more likely they were to debunk another's poor theory.

Late corpuscularist alchemy was the theory on which Boyle and Newton built modern physics and chemistry. From the quoted passages of Boyle and Newton that will follow, we will see that their arguments for the accuracy of the corpuscular hypothesis made liberal use of the Inference to the Best Explanation (IBE) discussed at length in Chapter 4. In short, they reasoned that they should infer the truth, or if you like a weaker appellation, credibility, of a hypothesis that best explains the range of relevant observational evidence. Boyle thought that the corpuscular nature of unobservable matter was most credible because it best explained the late alchemical effects of titration and precipitation, along with the behavior of gases, that were otherwise puzzling. As he put it, "One is never so sure to have resolved a mixture into its true constituents as when, with the same constituents, one has been able to recompose it."[11]

Newton inferred the plausibility of the corpuscular hypothesis from the fact that it best explains the universal application of mechanical laws to macroscopic and microscopic objects alike. While science typically moves forward by careful, incremental theory building, it sometimes lurches forward quickly and precipitously by these contingent hunches. These hunchy leaps forward are not uncommon in the history of science, and they vary in size.

In fact, my explanation for the rise of modern science is an IBE as well. I argue that the North's and West's historical dominance in science is best explained by the contingent and serendipitous cognitive leap to the modern corpuscular hypothesis. Once the atomistic hypothesis was sufficiently accurate—no longer hobbled by the vagueness of ancient formulations and the flabby mysticism of corpuscular alchemy—modern science could fully explore the reach of an accurate physics into the many dark corners of reality, from the chemistry of gases and the physics of combustion to the causes of chemical bonding.

As it turns out, many, if not most, scientific advances are the result of some contingency or other. A scientific discovery is contingent when it emerges or results from a process that isn't part of a rational procedure—a hunch, a religious epiphany, a dream, an accident of nature, a peculiarity of a warring culture, a chance meeting, a biological condition, a geographic boundary, and so on.

Serendipity is one kind of contingency, and stories about it in science delight. At the age of 19, Edward Jenner—the inventor of the smallpox vaccine—had a chance conversation with a former milkmaid who said that because she had had cowpox, she couldn't get smallpox. Alternative treatments were slower to develop, and so there is no telling how many more people would have died from smallpox were it not for Jenner's chance conversation.[12] Change the chance event in a given scenario, and important, life-saving ideas could go undiscovered.

Office assignments are not always made with research interests in mind. At Cambridge, Watson and Crick shared an office with Jerry Donohue, an X-ray crystallographer who had his own research program. Watson and Crick themselves noted the fortunate contingency of sharing an office with someone who was better able to represent possible DNA structures for them to consider. As Watson puts it: "If he [Donohue] had not been with us in Cambridge, I might still have been pumping for a like-with-like structure."[13]

This looks a lot less like a steady march and more of a lurchy advance. Once you recognize that there are other courses history might have taken, it is easier to appreciate the powerful role of contingency.

The Newtonian Revolution, so often presented as a carefully orchestrated movement, also began in contingency. Alchemy provided no sure characterization of the composition of matter. In the intellectual gumbo of the time, the causes behind the Newtonian Revolution almost *had to be* intellectually contingent, surprising, and unique. There was no scientific academy to coordinate vocabulary, to referee publications, or to cooperate across nations. In these environments, when a hunch hits the target, there may be no satisfying story to tell; no narrative that ties prior research to the target; no appeal to tacit knowledge, to subtle skill, to relevant expertise, and so on. The line of research stuck because the educated guess had latched on to the causal structure of the world.

Even though we have seen that the history is messy, many philosophers of science and epistemologists hope to extract clean lessons from the meteoric rise of modern science beginning in the time of Boyle and Newton. They propose to explain this progress by seeking answers to specific philosophical questions: did Boyle and Newton have knowledge of the microstructure of matter, and if so, when? When philosophers and historians of science insist on an answer, they are also insisting that the answers be cast in terms of familiar epistemological categories. Did Boyle and Newton *know* that the collisions among atoms in a gas were roughly elastic? When the method is a mixed success, and the procedure is a collection of variously guided guesses, posing questions like these seems like a hygienic practice, and attempting to answer them is a scholastic exercise.

These epistemologically based questions miss that there is a difference between understanding and merely being right. There is even a difference between *incomplete* understanding and just being right. Without implying that the act or process of understanding has clear identity conditions or forms a special epistemic category, these notions capture a phenomenon associated with the cognitive appreciation of the unobservable mechanisms that bring about an effect. Daniel Sennert could describe, in non-Aristotelian terms, why only the silver in an alloy of gold and silver dissolved in nitric acid could be reconstituted as pure silver. This moved many to speculate that the silver, throughout dissolving and reconstitution, retained its essential structure, because the silver disappeared and then reappeared (rather than being what Aristotelians called an imperfect mixture). Sennert's explanation was roughly right—that the essential structure was a corpuscular configuration that persisted through superficial transformations. But did he *know* this? Did he *understand* the nature of this dissolving and reconstitution? Did the term "corpuscle" *refer* to a specific thing in the real world?

The understanding of mechanisms, incomplete or full, gives you a fighting chance to produce theoretical success. That is why genuine understanding—the accurate cognitive appreciation of the nature of the causal mechanisms that bring about an effect—is positively related to theoretical success.

Philosophers interested in these cognitive achievements sometimes reject scientific realist interpretations of the history of science by appealing to problems with theories of understanding, knowledge, or reference used to make the case for scientific realism. My approach disciplines critics so that they must be modest about much of the specific machinery we have in hand, even if the elevated intellectual standing of science in many of 17th- and 18th-century Europe's urban centers is both powerful and undeniable. It may be a long time before we have a *theory* of knowledge, or a *theory* of reference, or a *theory* of understanding—let alone a comprehensive theory unifying them all—that makes sense of the myriad cognitive achievements and failures in the history of science. We enjoy cognitive achievements like modern chemistry, biology, and physics decades, and sometimes centuries, before philosophical ingenuity has worked out *theories* of those domains.

One thing that we do know is that much of the world we cannot see, hear, touch, smell, or taste. This part of the world is unobservable. But it is also where much of the action is. It is behind every flash of lightning and earthquake, every epidemic and germination. Knowledge about these unobservable items is called "theoretical." Before the 17th century, there was not a single theoretical insight that generated a good theory of matter.

Not all progress depends on having a theory of things unobservable, of course. Some progress is purely technological, embraced for its useful products, without a consideration of the unobservable mechanisms that make it possible. Quanats in 800 BC were engineering marvels that brought mountain water to the Persian plains.[14] Animal husbandry, dutifully practiced in 400 BC in Neolithic Greece, supplied a more stable food source than ever before. In fact, none of the familiar engineering or technological feats of the ancient world depended on having a good metaphysical theory of the causes of planetary movements, hydraulics, chemical processes, strength of materials, or geomagnetism—not the calendars in Babylon, the aqueducts in Rome, gun powder in China, the forged swords in Japan, copper smelting in 3500 BC, or the compass's use by 7th-century Chinese sailors. On the contrary, their metaphysical outlook was so profoundly mistaken that they were unable to build on it.

We now have a rough answer for what explains the rise of a much-improved metaphysical account of the world, as well as the speed of its ascent. The rise begins around 1600, when scientists at the border of alchemy, mathematics, celestial mechanics, practical medicine, and chemistry began to generate the right hunch about the physical structure of unobservable objects. A consensus emerged that most everything was made up of particles, some of them too small to see. It was this corpuscularian theory that got just enough right about the underlying nature of things. The theory performed well in other domains, as we would expect an accurate theory would. These two properties of corpuscularian theory—its approximate accuracy and performance—were enough to make it a strong candidate to give rise to the Steep Curve.

The Big Leap

This two-part explanation for rapid improvement of science in the early 1600s applies only to theories, not to mere instruments of calculation. To address the causes of the approximate accuracy and performance of the corpuscular theory—what we might think of as the correct metaphysical account of the unobservable world and its use as a theory—we need to distinguish two sources of intellectual progress in science. The first is formal, dealing with the method of investigation, and doesn't much depend on what the unobservable world is like. There were bursts of intellectual activity wherever a civilization flourished, and magnificent feats of mathematical and technological achievement. But none of this depended on having an accurate metaphysical worldview, a conception of what the unobservable world was like.

According to this simplistic story, mentioned earlier in the chapter, Isaac Newton emerges with a brilliant mind and a jaw set against the mystical, dark practices of alchemy. Officially opposed to speculation and hypothesis, he applied The Scientific Method to identify and describe the mechanical forces of nature, establishing through increasingly refined observations that the world in the small was like the world in the large. He ventured that the collisions between microparticles could be described by the same equations that governed the interactions of billiard balls and planets. A physicist no less than Ernst Mach said that not only did the Newtonian law of attraction best explain observed facts, but also it is "the only one that can be uniformly and universally applied to large and small bodies, to apples and to the moon. It is the only one, therefore, that it was reasonable for God to have adopted as a law of creation."[15] Mach was not the first to notice this unifying advantage of Newton's view. As early as 1796, Pierre-Simon Laplace declared that by following the assumption that chemical phenomena could be explained by short-range forces, "we shall be able to raise the physics of terrestrial bodies to the state of perfection to which celestial physics has been brought by the discovery of universal gravitation."[16]

It certainly is true that the application of particular celestial laws to microparticles in Newton's century marks a big break with the earlier alchemical tradition. The laws implicated causal properties of bodies, like mass and momentum, possessed by large and small objects alike. Alchemy and pre-Copernican astronomy hadn't offered any rigorously unified picture, though the grand and the small were sometimes said to be connected by mysterious associations of sympathetic magic.

So Newton should be given credit where it is due for risking the creation of a daring edifice. Though corpuscular alchemy anticipated several of the molecular properties that early chemistry would soon find—and this lubricated the conceptual path—the gap was huge. In the early 1600s, practitioners had fundamentally different conceptions of chemical combination, and none of them were much like Newton's. In the words of Newton scholar Dobbs:

> In the various forms in which corpuscularianism was revived in the seventeenth century, the problems remained and variants of ancient answers were redeployed. Descartes, for example, held that an external pressure from surrounding subtle matter just balanced the internal pressure of the coarser particles that constituted the cohesive body. Thus no special explanation for cohesion was required: the parts cohered simply because they were at rest close to each other in an equilibrated system. Gassendi's atoms, on the other hand, stuck together through the interlacing of antlers or hooks and claws, much as the atoms of Lucretius had before them. Charleton found not only hooks and claws but also the pressure of neighboring atoms and the absence of disturbing atoms necessary to account for cohesion. Francis Bacon introduced certain spirits or "pneumaticals" into his speculations. In a system reminiscent of the Stoics, those ancient critics of atomism, Bacon concluded that gross

> matter must be associated with active, shaping, material spirits, the spirits being responsible for the forms and qualities of tangible bodies, producing organized shapes, effecting digestion, assimilation, and so forth.[17]

There are large differences between these theories. Scientists have their own way of resolving these disputes, and then philosophers have their way of trying to understand the history of science. These ways are quite different, and it is no wonder that philosophers' intellectual fastidiousness find wanting scientists' practical eye toward confirming hypotheses. But the failure of practical science to meet the hygienic standard of philosophical scrutiny does not mean the failure of practical science to bridge any number of theoretical gaps like the ones seen previously. The theoretical gaps *were* bridged, philosophers be damned: the only question is how. All sequences of theory succession leave gaps; some sequences are gappier than others, or leave greater conceptual spaces than others. Scientists deal with all of these disputes in the same way they always have—by finding common methodological or theoretical standards that could persuade them of one side's case over the other.

As should be clear by now, the gap between precorpuscular alchemy and 17th-century atomism is bigger, and the border work less developed, than in similar cases. When historians and philosophers of science have finished tracing the connections between the launch and landing of a new theory and there is still a significant gap, it looks likely that the theoretical development was the result of a historical contingency. Perhaps a key scientist had a startling insight, or a poignant dream. Perhaps by chance two researchers begin working together who, it turns out, have complementary strengths. Whatever the case, it was not part of a rational plan, and there is no way to fill every gap with a story that makes every move forward rationally intelligible.

The weight of the historical evidence supports two jarring claims about the Boyle-Newton atomism that launched modern theoretical physics and chemistry. First, while Boyle and Newton weren't shamelessly guessing when they proposed that solids, gases, and liquids were composed of corpuscles, neither were they securely deriving that conviction from empirically demonstrated premises, any more than their alchemical predecessors were. Newton's speculations occurred in the Queries—a volume whose very title belies the idea that Newton was sure he was correct. The most comprehensive sweep of evidence provided only equivocal grounds for embracing atomism; they had the kind of supporting evidence characteristic of a good hunch. Second, in a sea of good hunches, this one stuck because it was accurate enough of the time for enough objects. This is not a clear-cut story we were told about the Newtonian Revolution.[18]

On the epistemic spectrum, Boyle's and Newton's key postulates were closer to fortunate guesses than bits of knowledge. Their inquiries did not entail that corpuscles were hard and spherical, or that their collisions were elastic. Instead, they were generating several presumptive judgments about what unseen causes best explain phenomena like dissolving, precipitation, heat, and barometric pressure, to name just a few. Whatever Newton's allegiance to alchemy, he advanced a bold chemical hypothesis in his *Opticks*, and he did it without the benefits of a well-tested prior theory. Query 31 opens with Newton reasoning that because there are many known attractive forces, other kinds of attractive forces may exist: "Have not the small Particles of Bodies certain Powers, Virtues, or Forces, by which they act at a distance," and doesn't the existence of Gravity, Magnetism, and Electricity make it "not improbable but that there may be more attractive Powers than these?"[19]

And if it is not improbable, Newton's intellectual excursions could form the basis of an account of chemical reactions. Focusing on those very local forces that might allow the smallest particles to aggregate to larger ones, Newton remarks in Query 31:

> The Attractions of Gravity, Magnetism, and Electricity, reach to very sensible distances, and so have been observed by Vulgar Eyes, and there may be others which reach to so small distances as hitherto escape Observation; and perhaps electrical Attraction may reach to such small distances, even without being excited by Friction.[20]

That is, even without the sort of excitation found in static electricity, there might be other causes of strong, local "bonding" of the smallest, insensible particles.

Newton continues to speculate, to guess and explore, as that is what the Queries were *for*. He was not positing his hypotheses as fact: they are quite literally questions. In fact, Newton's Queries found new theory by informed guesswork about what may or may not be the case. His choices represent contingency in action. Under the control of a uniform attractive force, the smallest particles could cohere and form aggregates whose "virtue" (or attraction) gradually decreased as the aggregates grew larger. Newton extended something like a gravitational physics to chemistry. His conjecture permitted chemists to interpret and measure affinities (or attractions) entirely in chemical terms:

> Now the smallest Particles of Matter may cohere by the strongest Attractions, and compose bigger Particles of weaker Virtue; and many of these may cohere and compose bigger Particles whose Virtue is still weaker, and so on for divers Successions, until the Progression end in the biggest Particles on which the Operations in Chymistry, and the Colours of natural Bodies depend, and which by cohering compose Bodies of a sensible Magnitude. If the Body is compact, and bends or yields inward to Pression without any sliding of its Parts, it is hard and elastick, returning to its Figure with a Force rising from the mutual Attraction of its Parts. If the Parts slide upon one another, the Body is malleable or soft. If they slip easily, and are of a fit Size to be agitated by Heat, and the Heat is big enough to keep them in Agitation, the Body is fluid; and if it be apt to stick to things, it is humid; and the Drops of every fluid affect a round Figure by the mutual Attraction of its Parts by Gravity.[21]

In just 130 years after the publication of Newton's *Principia*, his hunch had vaulted physics over a hopelessly mystical morass of alchemical folklore into theories of gases, chemical combination, and charges and electrical interactions.[22] It is hard to imagine a conception with greater power and machinery to unify a more diverse set of domains, from physics, chemistry, and areas of biology to celestial mechanics and geology, than Newtonian theory, broadly understood. And all of this successful unification by Newton occurred alongside an entourage of latter-day corpuscular alchemists who happened along his path, innocent of their poisonous solutions, mindless of the mechanisms of corpuscular attraction (like valence) and diffusion. Newton was right in ways that pushed the field forward, and wrong in ways that didn't prevent gross progress.

So Newton made a good guess, that the world in the small was like the world in the large. All is space and corpuscles. Newton correctly supposed that the motion of unobservables, like the motion of planets and projectiles, is space and corpuscles, space and celestial bodies, space and ballistics. Some might say Newton "guessed," and others might say that he "inferred the best explanation"; he inferred the hypothesis that, if true, best explained the range of evidence. But in this context of dawning science, the corpuscular view both vaults enormous territory and comes to rest on largely untested ground. Because of these two disparate properties of the corpuscular view, the threshold to maturity may be crossed suddenly.[23]

The Corpuscular Hypothesis: What Made This Hunch so Special?

Now let us take a step back from Newton's hunch. Why was the corpuscular hypothesis such an important development, and why was the theoretical assertion of the particle's hardness, roundness, and elastic collisions enough to provoke scientific developments that changed the world? There are three main reasons, and they are the three reasons we have seen for why most any theory thrives:

1. The theories encompass a wide range of objects that are central explainers of significant effects.
2. These properties unify objects in otherwise widely separated fields of inquiry. They are connected causally to those in other areas of physics, like

thermodynamics; to areas of pneumatic chemistry; to chemical taxonomy; and so forth.
3. Those assumptions are roughly accurate.[24]

Taken together, the corpuscular hypothesis—a hypothesis about unobservable states and processes—of the early 17th century was the first *theoretical* hypothesis to be both accurate *and* important (at least in the sense of instantiating reasons 1 and 2). Historians and philosophers of science sometimes goad and prick scientific realists for their theoretical progressivism by pointing to earlier incarnations of a theory that failed, or false theories that yielded useful results. They point to the fact that Newton's corpuscularism is half-baked (including a role for gravity but not valence), had poisonous ingredients (alchemy), or had patently false commitments (like static repulsion). These are all worth mentioning, but they don't spell trouble for a scientific realist.

The late alchemists handed Newton and others a corpuscular hypothesis that was getting close to the truth. What hunches following were so accurate that they then were able to drive early physics and chemistry forward? Measured by the standard of modern physical theories like thermodynamics and chemical theories of valence, a number of the Boyle-Newton hunches were not just accurate; they were correct about theoretical properties that were very important in characterizing the microstructure of nature. These correct hunches of the new corpuscularism, in particular, gave way to the Steep Curve in scientific progress. First, corpuscles are hard, and second, they are round. As Newton put it:

> All Bodies seem to be composed of hard Particles: For otherwise Fluids would not congeal; as Water, Oils, Vinegar,

> and Spirit or Oil of Vitriol do by freezing; Mercury by Fumes of Lead; Spirit of Nitre and Mercury, by dissolving the Mercury and evaporating the Flegm; Spirit of Wine and Spirit of Urine, by deflegming and mixing them; and Spirit of Urine and Spirit of Sal, by subliming them together to make Sal-ammoniac. Even the Rays of Light seem to be hard Bodies; for otherwise they would not retain different Properties in their different Sides. And therefore Hardness may be reckon'd the Property of all uncompounded Matter. At least, this seems to be as evident as the universal Impenetrability of Matter.[25]

In combination with Newton's second law of motion, corpuscles' hardness allows us to explain the macroscopic effects like pressure and temperature by appeal to their velocity, the elasticity of their interactions, and the density of the gas in the container. On Newton's view, the differently shaped corpuscles would bounce off of each other with trajectories that would be extremely difficult to predict. But macroscopic measures like pressure and temperature could still be calculated by averaging over the interactions of the differently shaped corpuscles. These are precisely specifiable features of microscopic properties and their mutual engagement.

Most important, there can be no question that these are hunches, not demonstrations or knowledge. And not only were these hunches by Newton, but also they were ones that had been percolating long before him in the shadowy history of alchemy.

Standing on the Unobservables of Giants

Newton hypothesized (or perhaps he would prefer "queried") that corpuscles are hard and impenetrable. Was there prior evidence supporting the view that there were unobservable particles with these properties? The answer is yes, and much of the evidence, as we saw, descended from Boyle, who in turn got it from Daniel Sennert.

Decades before Newton, Sennert put alchemical attachments to folksy tests. Around 1618, Sennert attempted to demonstrate atomism by passing a stream of alcohol vapor through a sheet of paper folded four times. He reasoned that the components of the vapor must be exceedingly small if they could travel through the invisible pores of paper so dense.[26] Because a range of postulated unobservables could have accounted for the observable effects of compounds dissolving in and precipitating out of a solution, it looks more as though the early alchemical postulate of fundamental constituents, or corpuscles, was just a good hunch, one that turned out to follow the causal roles we would later find in 19th-century chemistry.

To Sennert, breaking mixtures down and reassembling them—as you do when you precipitate components in a solution and then dissolve them again—formed a compelling case for atomism. On his view, we were limited by naked-eye observation, and probably not smart enough to know the mechanisms or forces that hold corpuscles together, but it was certain you could reduce any mixture into its original parts and put them together again.[27] And if this were true, how could the underlying reality of these processes be anything other than unobservable particles that had attractive and repulsive forces?

Historian of science William Newman traces a direct causal line from Sennert to Boyle and Newton:

> The hierarchical structure of corpuscles composed of smaller corpuscles [as posited by Sennert] would provide

> a fundamental building block to the mechanical philosophy of Boyle and would reappear in still recognizable form within the matter theory of his chymical heir, Isaac Newton. Underlying this belief was the clear conviction, inherited from medieval alchemy, that processes such as sublimation, calcinations, and dissolution in corrosives provided ocular testimony to the analysis of matter into its more fundamental corpuscular constituents.[28]

That said, Sennert, Boyle, and Newton offered no explicitly "mechanical" explanation of these processes by appeal to the size, shape, and arrangement of atoms. But Sennert and Newton both thought that the corpuscles interacted mechanically, had "elective affinities," and had "attractive powers" to account for the strange affiliations some substances form for others.[29]

The hypothesis that it was unobservable corpuscles that produced these effects had some currency. But many quite exotic visions of these corpuscles were compatible with what they observed. Descartes (1596–1650) ventured a similarly mechanistic explanation, in which the air was composed of particles in "restless agitation" with each other. Each corpuscle tried to fight off all others coming close to it, and Boyle compared Descartes's explanation to his own. In his essay "Spring of the Air," Boyle compared air—whose "particles are very small"—to a piece of wool fleece that has been compressed. It is always trying to expand, and can do so when the compressive force is eliminated. The key concepts here are similar to the modern concept of the kinetic theory of gases. In his essay, Boyle did not choose one model over another, but he did think his own was easier to understand.

In the end, of course, the alchemists were proven wrong. Acidic solutions didn't have their effects by their particles stabbing our tongues. And alum was not octahedral, nor did it fit into a matching vacuum in a solution. As later alchemists pulled away from these mistaken conceptions, they made some guesses—guesses that didn't have to be perfect to push us forward. Success rests not with perfection but with decent accuracy about enough of the dimensions of (in this case) corpuscles. You can't expect that kind of luck. If you can get it, you want a good enough theory tested with a good enough method.

So to summarize: we should get clear about what this explanation for scientific progress entails and what it doesn't. To be used to excellent theoretical effect, the early 17th-century corpuscular hypothesis only has to be actually *true enough*—it doesn't have to be predictable from earlier theories or even believed by experimenters. It doesn't have to be the best theory imaginable, or even the best theory available. It doesn't have to prove itself better than theories that existed but were never tested. It doesn't have to have outcompeted all other potential or credible rivals—it only has to, in fact, get entrenched. It only has to be good enough, true enough, and important enough.

Dispelling Myths about the Steep Curve

The Newtonian hunch turned into a significant advance, but it didn't arise from the secure, tried-and-true application of the scientific method. On its own, the scientific method is powerless to improve science until the theories it is applied to are good enough. Just as the scientific method can't improve specific astrological theories of personality, alchemical theories couldn't be improved simply by applying experimental methods: there needs to be something more. Consider van Helmont, a key figure poised on the cusp of Boyle's emerging corpuscularism and to whom Boyle deferred. In that environment fertile for progress, even the remnants of Aristotelianism could be bound up with views that lead to technical innovation. Thus, van Helmont's "alchemical doctrine," that water is the unique primordial substance, inspired a new experimental technique using "bell jars." These containers, used to receive and identify the gases given off by reactions, would become part of laboratory equipment of both alchemists and chemists in the mid-17th century. Given practical applications such as this, it is not surprising that alchemy was slow to die. Around 1670, Boyle was still breaking loose from the alchemical principles of Sympathetic Magic, soon to give way to chemistry's laws of affinity and attraction. But as late as 1650, van Helmont contended that, if someone defecated on your doorstep, hours later you could slap a hot iron to it and burn their anus by the mechanism of "dorsal magnetism."[30] Alas, there is no such mechanism, and yet Boyle's *Sceptical Chymist* had van Helmont as its chief conceptual inspiration.[31]

There is no evidence that Boyle and Newton kept their irons hot for just such an occasion. But their indebtedness to alchemy was evident in their behavior and in their professed allegiance.[32] So there is also no way of extricating Boyle and Newton from this concoction of occult beliefs. When you have a potentially good theory laced with myths of transubstantiation, sympathetic connections to alchemical qualities, and numerological gymnastics, it is on the threshold of reliability. But theories on the brink warrant diagnoses as uncertain as their condition. It doesn't make much sense to ask whether the success of Newtonian corpuscularism is owed to the reliability of scientific method when the reliability of scientific methodology itself depends on the accuracy of Newtonian corpuscularism. They are jointly implicated in success. Focusing just on the assessment of scientific methodology, then, we can say that, until Newtonian theory was sufficiently developed, applications of scientific methods of reasoning would not be, and were not, reliable.

Of course, my view of scientific progress still celebrates the power of the scientific method when applied to a good theory. It does not undercut my view to suppose that scientists should be taught (and perhaps a few are taught) to look for counterexamples to theoretically expected outcomes and search for defeaters of a favored hypothesis, and not to rest comfortably with the first explanation that "feels right." You never know when a well-placed experiment or application of the scientific method will confirm an observational hypothesis of practical medicine, for example, as it did in ancient Indian treatments against viral exposure. If these are concessions to an opposing view, at least they are concessions to virtue. At the same time, I have argued throughout that scientific method only works well when you have a good enough background theory. And so, with Kuhn, I agree that you can't explain changes in background theory as the result of the scientific method. The range of counterexamples and defeaters we entertain are often remarkably closely anchored to the existing theory, and when that theory is poor, we are not purchasing theoretical refinement by considering, say, artifacts that a poor theory tells us to control for. That's why contingency matters. It opens up the range of independent forces that operate on our naturally conservative hypothesizing.

Two sources of information contribute to the quality of such scientific reasoning: the reasoning method, such as *modus ponens* in deduction or multiple regression in statistics, and the theory that the method is applied to, such as modern optics, molecular genetics, or thermodynamics. During these episodes, it is easy to resort to simple narratives, say, that the mechanism of Descartes, Newton, or Boyle was arrived at by a rational scientific method. This impression seems undeniable, because the only alternative explanation for their accuracy is a good theory, which actually hadn't been proposed yet and wasn't available at the time. On the other hand, it is hard to see how the scientific method could have truth-generating power without applying it to relevantly correct theories.[33]

To take just one example, we may ask: was Boyle's theory reliably produced, created by a method or mechanism that tends to be truth yielding? If so, can we say the same of van Helmont's, Sennert's, or Starkey's earlier, less accurate theories? This is a difficult question to answer simply and directly, precisely because the reliability of the method used to study chemical composition—its ability to produce approximately true beliefs about composition—depended on the accuracy of the chemical theories that were embraced. Reliability demands only that its origin be accurate, not whether that accuracy grew incrementally from atheoretical experimentation, or suddenly from a good guess. In one way, then, no, Boyle's theory was not reliably produced, because while Boyle's theory is decent, the immediately preceding theories used to confirm Boyle's theory were still deeply problematic in ways that Boyle's wasn't. To put it somewhat paradoxically, the scientific reliability of chemical methodology used to support Boyle's theory was established only through the adoption of his theory.

At the same time, it is tempting to answer that yes, Boyle's theory was reliably produced, because the mechanistic suspicions of "occult powers" and of the Aristotelian theory of forms helped to produce modern corpuscularism—and such suspicions in fact constituted a reliable belief-forming principle. What can we say of the forming of those valuable suspicions? Were *they* reliably produced? The answer is naturally indeterminate. It is customary to say that Boyle knew that matter was composed of small particles called corpuscles, and so did Sennert several decades before him. But it is also natural to say that Boyle and his immediate predecessors didn't embrace the atomic theory of matter; they really didn't know that the atomic theory of matter was true.

We are therefore faced with a difficult choice. People interested in the nature of knowledge, particularly scientific knowledge, might ask whether 17th-century mechanists "knew" that the corpuscular theory of matter they advocated was (approximately) true. If you say yes, then it is difficult to explain their behavior, because that would mean they were aware of the many phenomena they wanted to explain but could not. If you say that what they had was not knowledge, then it is hard to explain how this information allowed such accurate predictions about the behavior of solutions, and explanations for why solutions that could be poured through filter paper without residue could, by precipitation, reassemble the metals dissolved in the solution.

On the other hand, people devoted to the reliability of the scientific method might ask whether the late 16th- and early 17th-century founders of modern physics and chemistry invented what we now call "the scientific method." If we say yes, it is not clear what we would call the same set of principles of experimental design that guided 11th-century Islamic inquiry into the nature of light, or Grosseteste's early 13th-century method of "resolution and composition." If we say no, we might wonder why science did not enjoy a steep rise even earlier, when the scientific method was "invented."

The impulse should be to follow the evidence even when its complexities defy simple dimensional judgments; it should not be to come down hard on one side when the evidence won't take us there. At most, we can say that corpuscularism of Boyle's sort was in significant respects true, and its accuracy along those dimensions was crucial in producing scientific knowledge in chemistry. Simply put, the evidence for corpuscularism was not powerful. But what the evidence supports, and all it supports, is the claim that the corpuscular alchemy of Boyle and his immediate predecessors was *good enough* to ground an inference from small, hard corpuscles to the performance of gases and the laws (like Boyle's Law) that link them.[34]

Thus, the answer to these questions regarding method and knowledge is this: the best explanation for the dominance of this brand of Enlightenment Science is that Boyle and Newton had a theoretical hunch, and it was roughly accurate enough that it persisted, even flourished, under diverse tests. *When a theoretical hunch stands magnificently against the pushes of history, psychology, and culture, there are few other explanations for its triumph but improbable coincidence or fortunate truth. The best explanation for how a hypothesis overcomes psychologically idiosyncratic fluency, culturally parochial commitment, and political ideology is that the guess is accurate.* Any less than that and the explanation likely fails to account for the effect.

We should dispatch with a handful of modern myths that make up the lore of the Newtonian Revolution. The first myth is that Newton was a distinctly modern figure and a dogged opponent of alchemy—that he pitted the enlightened practices of early chemistry against the dark forces of alchemy. On the contrary (and as most academics have now heard), Newton himself did serious work in alchemy. And Boyle and Newton both expressed a mixture of admiration and contempt for alchemy.[35] At the same time, most of the alchemical theories known to Newton and Boyle were shrouded in mystery, and were not clear enough to document their connection to the corpuscular theory. Not surprisingly, there is no evidence that Newton worked out the corpuscular theory to repair a bad theory, to resuscitate an old one, or to extend an existing one.

The second myth is that the rise of atheism was behind the rise of atomism. Whatever the actual effects of atomism's severe ontological standards, this myth makes bad biography. Indeed, like their philosophical contemporary, John Locke, Newton and Boyle were champions of transubstantiation; they apparently believed that, during mass, bread and wine were transformed into the body and blood of Jesus. Their religious standing, as well as their patronage, depended on a gentle hearing for this doctrine. Their religious conviction was threatened by the easy observation that the Eucharist undergoes no perceptible change during the mass, and this led Newton and Boyle to suggest that while the nominal (superficial, observable) essence of the Eucharist remains the same, its real essence changes. This religious belief, in fact, may have lubricated the path to Newtonianism, because it allows believers in transubstantiation to treat the Eucharist as having a real essence that changes (at the unobservable level) despite no observable differences in the Eucharist during the mass. The prominence of the church, both Catholic and Anglican, gave the alchemist's rhetoric more currency.

The third myth is that Newtonian theory presented in the *Principia* in 1688 could fall from the pages relatively unrevised. While that was true of certain equations, some of Newton's thermodynamic hunches concerning the nature of corpuscles and their interaction were flatly mistaken. For example, Newton ventured that pressure in a closed container results from static repulsion between molecules, not from collisions between molecules moving at different velocities:

> If a fluid be composed of particles fleeing from each other, and the density be as the compression, the centrifugal forces of the particles will be inversely proportional to the distances of their centres. And, conversely, particles fleeing from each other, with forces that are inversely proportional to the distances of their centres, compose an elastic fluid, whose density is as the compression.[36]

Newton's wake left no guidance in the matter of the precise intrinsic character of the particles that made up gases, liquid, and solids. Yet, Newton's hunch broke toward progress despite these errors, because they were dominated by the accurate depictions that were dramatically important to the development of the ensuing chemistry.

The fourth and final myth is one that has been already approached in this book: that the dawn of The Experimental Method accounts for the success of Newtonian theory. Though the myth persists in college classrooms that Francis Bacon was the father of the experimental method, the first to develop and apply principles that involved holding some conditions constant and manipulating others to record predicted outcomes, Bacon was a celebrated codifier but late observer of the scientific method. The Bacon myth was a convenient story to launch the grander narrative of scientific growth during the European Enlightenment. The experimental method had been used for at least 600 years by then, and its effect depended entirely on the initial quality of the theories being tested. Before Bacon and Newton, the experimental method was applied with only modest effects, but it was unquestionably used. In 1025, Ibn Sina's (Avicenna) description of the methods of agreement, difference, and concomitant variation, all found 800 years later in John Stuart Mill and central to all versions of inductive logic and the scientific method, were used to test medical conjectures. Seven centuries earlier, Islamic scientists began developing what we now recognize as the modern canon of experimental method. As early as 800 AD, the Muslim chemist Jabir ibn Hayyan (also known by the Latin name, "Geber") introduced controlled experiments into chemistry. Around 850 AD, another Muslim polymath, Al Kindi, experimented with plants to produce fragrances, and also advanced an alternative to Euclid's extramission theory of light. Shortly after, Al-Razi (850–923), a Baghdadi alchemist and physician sometimes called "Rhazes," penned *Secret of Secrets*, a working text of chemical practice. An expert at distilling liquids, as well as classifying substances into metals, boraxes, stones, salts, and vitriols, Al-Razi used taste and solubility as a basis of classification. Only 100 years after al-Razi's death, a torrent of Islamic scientists practiced and articulated a recognizably modern scientific method. In 1021, Alhazen (Ibn Al Haytham) nailed down an intromission theory of vision and produced a landmark work in optics. Three centuries later, Al-Farisi used a camera obscura to show how images become more sharply focused as apertures get smaller. (He also noted that they invert and shift left to right.) Al-Beruni's (973–1048) achievements are too numerous to summarize both graciously and briefly, but they include astronomical investigations that corrected Ptolemy, methods for locating the planets (a technique called "three point observation" used later by Brahe and Copernicus) and gauging their acceleration, and a hypothesis that light is faster than sound, to mention just a few advanced insights.[37]

Champions of the experimental method hope, of course, that a reliable method will compensate for the unevenness of human talent, variability of individuals' knowledge, and jagged and unpredictable workings of historical and cultural accident. The empiricist Pavlov's methodological dictum was offered instead as a mortal's only hope: "With a good method, even a rather untalented person can accomplish much."[38] In fact, the father of classical conditioning went so far as to claim that "Everything is in the method, in the chances of attaining a steadfast, lasting truth."[39] This sentiment has dominated methodological reflection on science in the 20th century.

Now standing at the summit of 21st-century science, scholars and scientists in Europe and the United States regularly conclude that, even with the experimental method, "science flourishes only in liberal-democratic environments."[40] We long to make sense of the forward motion of what sometimes seems this tireless beast of science, and the standard explanation of this progress sounds pretty airtight. The story we love to tell is that Galileo and Descartes, Boyle and Newton, all resuscitated a kind of mechanism and atomism. Boyle paid special attention to gases, Newton to solids that were macroscopic. They both went so far as to hypothesize that the motion of corpuscles caused the behavior of matter, despite Newton's assurance that he ventured no "hypotheses." Using the experimental method that descended from Bacon, they recruited tools of the day, like the air pump, to rigorously test molecular hypotheses. After that, it was off to the races. The measurement of molecular properties like elasticity and sphericity launched quantitative chemistry. Theory blossomed under the light of free and open inquiry where, for the first time, scientific societies convened and members exchanged ideas, false views could be corrected under the strain of public discussion, and truth—not ideology or authority—was the arbiter of belief.

But this airtight story leaks, and not just for the fussy reasons historians cite. True, Newton had another life in alchemy, he believed in transubstantiation, and he thought that the air pump performance resulted from the static repulsion of particles. But those are minimal compared to the fact that the "airtight story" is false to the core.

CONCLUSION: ACCIDENTS, CONTINGENCIES, AND SCIENTIFIC REALISM

The contingency of scientific progress complicates the story of the scientific progress we like to tell, but it doesn't sink it. Scientific realists famously argued that the truth of our best theories is the only explanation that doesn't make the success of science a miracle. Van Fraassen offered a Darwinian reply: [T]he success of current scientific theories is no miracle. ... Only the successful theories survive—the ones which *in fact* latched on to actual regularities in nature."[41] But van Fraassen is too modest. The steep curve is explained by latching on to actual regularities in nature, sure enough. But there is latching on not just to the observable regularities in nature, but to the deep, unobservable causes in nature—like atoms and molecules.

In short, the best explanation for the rise of the modern science of the Enlightenment (based on the most plausible inductive inference) is that a confluence of lucky events (rather than deliberate applications of innovation in the experimental method) caused the Scientific Revolution in the 17th century, a story quite different from that told in textbooks and by philosophers. Alert readers may wonder how a powerful upward thrust could be produced from a corpuscular hunch—so central to the Newtonian view—that has been rendered false by the rise of quantum mechanics. It is one thing to endorse a best explanation when it is true, quite another when it is false. After all, the indisputable success of quantum mechanics is sound reason to reject the idea of corpuscularism, at least in its Newtonian forms. But my argument for realism requires no reinterpretation for the contingent rise of quantum mechanics. The same story can be told in quantum mechanical form, in which corpuscles are independent entities with large enough quantum numbers, whether the corpuscles are segments of the field in which the energy density is high, points of force in a field, or separable items with independent dispositions.

Following the most recent ice age, there have been bursts of intellectual activity wherever a civilization flourished, and magnificent feats of mathematical and technological achievement. But none of these practical feats depended on having an accurate metaphysical worldview, a conception of what the unobservable world was like. This is what Boyle and Newton supplied. Though instrumental and mathematical sciences populated the premodern landscape, theoretical science did not take off until natural philosophers and scientists got it right about the nature of unobservable phenomena. And once it took off, science never turned back. Now, the sense of understanding is a reliable cue of approximate truth, at least about the basic issues in the field. The asymmetric nature of causes captured by hypotheses give theories their sharp, recurved teeth, and they don't easily release their catch, even when their initial bite was a lucky one.

7

CONCLUSION

PEOPLE FIND WONDER IN MANY things—an endlessly clear night sky, the vastness of an ocean, the eerily familiar gaze of a primate, or the delicate choreography of a chemical reaction. It is hard to imagine this experience of wonder without also aching to understand. So we reach for answers, and from the sense of understanding we get, we cobble together the best explanation our time and training allow.

We arrive at a good theory by any means, whether by chance, by grinding experiment, or by any combination in between. As we have seen in this book, sometimes this sense of understanding leads us to the truth, and sometimes it leads us astray. A lucky hunch can vault us forward, until we are far ahead of the experimental evidence of the time. Once a coherent body of lore is assembled, approximately true theories support inferences to the best explanation, and then drive us ever forward.

But as we have also seen, the system is far from perfect. When the web of beliefs that supplies fluency is a bad one, like the theory that radiation is a topical contaminant that can be scrubbed off or that climate change is unaffected by human presence or products, people holding such a view get a sense of understanding from assertions that are false, sometimes profoundly so.

And now, in the present day, having access to the greatest science and creating the greatest infrastructure the world has ever known, the US government still uses science to make policy as though we all live in a medieval village. The methods still allow individuals to rely on their untutored and egocentric sense of understanding, never a good recipe for truth-tracking pursuits. For just one example of many: on March 26, 2014, the House Committee on Science, Space, and Technology, one of the most influential science policy bodies in the country, met with the US president's chief science advisor. Members of the House Committee complained, in effect, that current science-based initiatives to slow global warming uncomfortably conflicted with their sense of understanding. As one member put it, it just didn't make sense to say that people were causing global warming. After all, the earth had warmed up between the ice ages when there were no people around, so how could we blame humans for the current warming? "Just because we're alive now," he reasoned, "the tectonic plate shifts aren't gonna stop, the hurricanes [and] tsunamis aren't gonna stop, the asteroid strikes aren't gonna stop."[1]

Unfortunately, it is far from uncommon for House members, Senators, and most other politically committed people to have no difficulty substituting expert scientific judgment with their own. Their sense of understanding tells them what a good scientific research program looks like, and what kinds of research programs are likely to be best for the country. But that doesn't mean that they're right.

Once you are psychologically steeped in a *good* theory, the sense that it makes can then drive better decisions and correctly reject the dysfluent and recalcitrant approaches that have been proven to be worse. Our sense of understanding can lead to good policy, then, when our best science orients it, and our best science is good enough. Fluoridated water, vaccination, antismoking campaigns, response to famine, limits on radiation exposure—these evidence-based innovations have vastly improved our health and reduced our suffering. It is true that some people will always have some strong convictions that don't enjoy evidential support, and we may never prevent them from expressing their strongest convictions, however misguided. But we can do our best to rob these convictions of their influence by starving them of nutrients.

How can we do this? While democracy has often gotten it wrong when using science, there are ways to ensure that it goes right (or at least not totally off the rails). To conclude this work, I will explore how to keep our policy attention firmly focused on scientific evidence, even when false, hysterical, or corrupt influences on the sense of understanding threaten our well-being.

The modern conception of democracy came into the world with an 18th-century vision of human cognition and well-being. The assumptions of early democracy held that people had vastly unequal intellectual skills that sorted them for citizenship and capabilities for, and rights to, well-being. This psychological theory, the outcome of casual observation and parochial prejudice, endorsed slavery, elevated to law a contempt for women, and abandoned the impoverished to uncertain charity. And it assumed that we could do no better than to have stakeholders argue in a public forum about what policies to live by. Just as certainly as we now know that early democracy was wrong on slavery, women, and the impoverished, we also know that the nonexpert electorate fighting over which side of an issue can be louder does not result in the best policies to promote well-being. A modern democracy must recast policymaking in a scientific image. Without scientific expertise informing the policy discussion, it is rarely the case that casual stakeholders are able to separate the wheat from the chaff.

Today, our psychological science is more sophisticated, but the psychology of governance has not much changed. People still accept inaccurate policy explanations and justifications because they feel right—they activate that flimsy sense of understanding. And new technologies provide ever more ways to tap into that sense of understanding. For example, explanations of human behavior rooted in cognitive neuroscience induce this glow of recognition and convey the blissful sense of understanding. Thanks to heavy media attention on these new tools that can peer into the inner workings of our brains, the lay public can now imagine Broca's area lighting up as we work to find the right word, as well as envision dark expanses in the prefrontal cortex of a remorseless criminal. News media report that scientists can read a magnetic resonance imaging scan of the visual system and identify the scene that the subject has just viewed, highlighting the vivid prospects for transforming daily life. We are then invited to speculate whether a computer could read a tetraplegic's brain scan and execute their commands, and whether scan reading could provide the blueprint for all-purpose brain-reading devices that escort us through a person's visual experiences, or even their dreams. These tidy, intuitive, and vividly imaginable findings of cognitive neuroscience are gripping. People like this feeling of fluency, however counterfeit the product, because they think they can "see" the place in the brain where actions and ideas have their root.

But these tools also lend themselves to more disturbing uses. A new breed of businesspeople calling themselves "neuro-entrepreneurs" claim almost unimaginable power to peer into the minds of consumers or voters, to give them what it has been determined they want or to tell them what they need. And citizens believe what they're told. Nielsen NeuroFocus[2] and other "neuromarketers" trumpet unguarded stories of triumph over consumer resistance or the crafty disabling of voter deliberation and discipline. A few well-placed references to "a squirt of dopamine" and "activation in the anterior prefrontal cortex" and you can just *feel* the power and profits crest. The public's neurophilia is worth studying in part because it may be exploited—prematurely, perhaps even cynically—by public-policy entrepreneurs attempting to shape cultural tastes and political convictions.

But neuromarketing is not the singular face of science in politics; it is just one influence of many on citizens' judgment. And we can decide to regulate these influences if we think regulation would provide a surer path to just policy and human well-being.[3] The decision about whether to use science in policymaking should not depend on the financial motives or poor behavior of advertisers and consultants; it should depend on having a clear understanding of which aspects of science, when applied, disable people's judgment and compromise their freedom. And conversely, we have to know if, how, and when people understand scientific explanations and incorporate them into their judgments. Scientific literacy, it seems, would be crucial for true self-governance.

Unfortunately, given the evidence, there is little reason to suppose that a national push for higher science standards in primary and secondary school will accomplish the necessary science literacy for responsible self-government. This competence must arrive by another route: a science of policymaking, to which citizens can contribute. This sort of policymaking would rely on mature science we know to be accurate, not intuitions regarding what we think is true. In the near future, issues of great human significance will come before the American electorate—stem cell therapies, genetic engineering, and imaging techniques bearing on legal competence, to name just a few. And only a process of policymaking that defers to our best scientific explanations—tested by experts and vindicated by their success—can secure scientifically responsible policies governing these issues.

Though we have much improved policies to prevent and manage famines, we still fall far short on a number of other major policy fronts. We have no responsible, effective response to genocide and poverty across the globe, or to poverty and poor public education domestically. The very science that uncovered the structure of motivated, and sometimes inaccurate, reasoning may be called upon again to guide us through this self-deception to the land of good policy. But ours is a world where arriving at the right solutions demands accurate theories, and getting that could require that we have the right explanations. Given that logical chain and the story told in this book, we should expect that, like good science itself, good policy often depends on historical contingency. We won't always see it coming, and we can't always make it happen. But two things are certain: sometimes we can improve policymaking (because we have), and we will always try to explain it when it improves.

Explanation is a natural activity, like gossip, problem solving, and cooperation (though Hobbes may disagree on the last point). As a natural process, explanation can be studied scientifically. This book was designed to develop a better understanding of the psychological principles that guide judgment. The first order of business has been to see how people actually go about explaining. Do people search for certain kinds of explanations over others—those invoking deep causes over observable ones, mechanisms over functions, simpler over more complex? A growing number of gifted psychologists have taken up this question: Susan Carey, Alison Gopnik, Frank Keil, and Tanya Lombrozo, to name just a few.[4] Their results have shown that psychological features of an explanation exert a powerful influence on people's judgments and acceptance of new beliefs. The new psychological study of explanation promises not just to help us understand ourselves better, but to create routes to human well-being through the political process.

From the many cognitive processes that form explanations, psychologists have traced a variety of cognitive biases or deficiencies in explanation, including one called the "illusion of explanatory depth." Frank Keil and his colleagues have examined this tendency. When they are first asked, people dramatically overestimate how much they understand about how an object works.[5] They believe they understand in far greater detail than they really do about how some aspect of the world works and why some given pattern in the world exists.[6] This illusion is demonstrated by asking people to rate how well they understand devices such as helicopters, cylinder locks, and zippers. Once they give their self-ratings, they are asked to describe in detail everything they know about how selected items work. Almost invariably, the subjects fall far short of the expertise they believed they had and struggle to form a coherent explanation for how helicopters fly, locks lock, and zippers zip.

Struggling to come up with a description has an interesting effect. When asked to rerate their understanding (1) in light of their past explanation, (2) having tried to answer a vital diagnostic question (e.g., How do you pick a lock?), and (3) having been given a brief expert explanation of how the device works, the subjects grow less confident. So you may begin this task thinking that you know how a lock works, but you realize you're less able to describe how a lock works than you might have thought just a few minutes ago.

There has been no hesitation in public policy research to take seriously the aforementioned findings of the psychology of explanation. Policies typically are complicated; they have many moving parts, and people tend to think they understand them far less than they actually do. For example, starting with Frank Keil's research on illusion of explanatory depth (IOED), Philip Fernbach and his fellow researchers proposed policies whose implementation and consequences their subjects don't understand.[7] To reveal this lack of understanding, all you need to do is ask them to explain how a particular aspect of the policy would work, or what steps would need to be taken once the policy was implemented. All of a sudden, as with Keil's experiments, people find they don't understand their preferred policy. But rather than give it up, they moderate their preference, moving it away from the acknowledged extremes, to shield themselves from the natural demand to explain the mechanisms of a more extreme view in more detail. In short, the experience of dumbfounding—in this case, being unable to explain the mechanisms responsible for their policy preference—undermines their sense of understanding. Accordingly, they moderate their views so their extremity doesn't provoke a demand to explain them. People obviously feel strongly about their political convictions, but the lesson contained in this research is sobering. If extreme political views are greeted with higher evidential standards, then many impassioned people will withhold their genuine views in favor of less contentious ones, in anticipation of being unable to support their views by detailed appeal to specific mechanisms.

The exploration of the psychology of explanation has much to offer here. We need to better understand why we take the unreliable cue of sense of understanding and then reject competing hypotheses for being outlandish when they conflict with it. Understanding this phenomenon would be reason enough to study it, but we also persist in cruel and arbitrary policies owing to their cultivated fluency. Explanations for why an alternative policy might be superior are too often prematurely rejected in favor of the one already entrenched in our minds: why are lawmakers in the United States, the wealthiest of nations, unable or unwilling to move its child poverty rate out of the bottom quartile of industrialized nations, to create a comprehensive policy on immigration, or to resist making arcane judgments in ignorance rather than defer to genuine experts? Surely sometimes the cynical explanations are the most accurate, ones that involve moral failings like willful stupidity, indifference when suffering is kept at arm's length, their biggest campaign funders or lobbyists telling them how to vote, their cultural commitments preventing them from taking seriously certain kinds of suffering, and so forth. But just as certainly, sometimes it is cognitive error that constitutes one more kind of bad reason to accept an explanation. Empirical work on explanation, elucidated in this book, can uncover these treacherous sources of influence and persuasion and apply them to the realm of public policy.

Scientific realism acknowledges human limits, but values testing them. Public policy, too, can overcome cognitive limitations, but it first must acknowledge that we have them. The limits on memory and attention described in Chapter 2 drive institutional arrangements designed to free us from those bonds. Automatic investment to combat undersaving, crafting choice sets to enhance decision satisfaction, and subsidizing leisure activities that improve health (like exercise and the arts), to name a few, improve human well-being, a metaphysical fact about us that is independent of whether we have theoretical knowledge of these processes. At present, our democracy allows, and sometimes encourages, choices that we have weighty scientific reasons to believe will make people's lives worse. Like children learning to ride a bike, we know they are going to hurt themselves, but we let them make the mistake. After all, how else can they learn unless we let them suffer the unambiguous feedback of painful error? Well, it turns out learning is complicated, and in the realm of policymaking, much of it can be achieved without risking cancer, child poverty, or loneliness. If we are impressed by our theoretical knowledge in the physical and biological sciences, it is exciting that we are now approaching similarly advanced psychological and social knowledge of the sources of well-being.[8] Some object to the use of policies that embody this knowledge to improve well-being, arguing that we then are not making our own choices. But we have to remember that we did not autonomously choose our existing policies either, and yet our biases toward the status quo give a pass to our default policies. It is hard to imagine people now objecting to our reliance on fluoridated water, seatbelts, and radiation regulation, especially if their reason is that they should not be made to do what science shows is best for them. And yet, this is the looming objection made to well-being policies based on psychological and social research. Ignoring that knowledge and yielding to antiquated tropes about freedom undermines human autonomy. We should prevent our democracy from patronizing citizens who think they most enjoy liberty when they are free to act in deepest ignorance. Champions of democracy, too, suffer from false climbs.

These are all reasons for scholars and laypeople alike to take seriously the new psychology of explanation. But there are also themes in the book directed at trained philosophers. If there are both cultural and environmental Umwelts, the contemporary philosopher's Umwelt rewards participation in protracted disputes kept alive by philosophers' appetite for vivid and outsized attachments to abstract problems. Inference to the Best Explanation (IBE) is one such problem. There seems to be no scholarly consensus about its reliability as an inference rule or the severity of the epistemic standards it must meet. But, with few exceptions, philosophers of science agree that scientists use it, so critics of IBE are left to argue that scientists are wrong in their use. Of course, philosophers are free to argue whatever they want, up to and including that IBE is a deeply defective pattern of inductive inference and that scientists ought not use it. But given the intellectual distinction earned by the core theories of modern science, from biology to physics, that has been a difficult reform to sell. If scientific explanation serves the intellectual goal of seeking truth, and scientists use IBE, it's hard to make the argument that IBE is intellectually suspect, is simply circular, or is inferior to philosophical alternatives that have not yet been named. You could prolong the dispute, I suppose, but if IBE resulted in progress, it could only be for reasons that science unveiled. IBE works, so it is unclear the ultimate purpose of philosophers' criticisms of IBE. Once IBE latches on to the truth, there is no turning back from the scientific progress that comes with it.

The ontic account of explanation I advance in this book helps to clear up this confusion over the apparent circularity of IBE. It distinguishes the metaphysical fact that the flipped coin under my hand is either heads or tails from the epistemological fact that until I uncover my hand and look, we won't know which it is. It separates the epistemic fact that our inductive evidence is always incomplete from the metaphysical fact that the truth status of the theory is independent of our evidence for it. Scientific progress occurs when we can bridge that gap.

Once we have a good theory, its advance is a one-way ratchet. There isn't much evidence of an alternative. Do we have good examples of institutional commitment to thermodynamics being withdrawn, only to teach another approach? These cases are hard to find. Mature sciences do not easily surrender their achievements, and regresses are uncommon. The purported cases would be strained. You might propose a relatively small, problematic pocket of an otherwise excellent theory. A test might be lax, for example. We may be ignorant of our errors, or we may lack the imagination to consider unconceived or empirically equivalent alternatives to the dominant theory. But this is not the explanation I propose. Scientific progress is a one-way ratchet because our best theories are approximately true. Those theories describe causal features of the world that, once engaged with other theories, don't let us regress. How, exactly, could our current physical theory of nuclear radiation regress so that we can no longer explain the effects of radiation in terms ionizing atoms but still explain the damage that radiation does to the human body? Simply supposing that we could explain it seems curiously hopeful. And could we then use the resulting new conception of radiation to explain the apparent accuracy of radiometric dating? On the other hand, trying to explain the damaging effects of radiation without appeal to ionizing atoms leads to endless inconsistencies or causal impossibilities. There is no way to carry out these revisions piecemeal. And because these revisions in the basics of the theory can't be contained, there is no way to make them and still maintain the intellectual integrity of related sciences and their products.

Philosophical theories of explanation have always sought to provide normative guidance about how a good explanation ought to be constructed. It is not always clear whether these standards come from common sense or from some deeply held philosophical conviction about what is required to render an occurrence unmysterious. Recent work on the psychology of explanation shows that some of the normative features in philosophical theories of explanation are expressed in our psychological preferences of explanation: we often prefer simple over more complicated explanations, and we prefer appeals to deep causes unless probabilities seem very illuminating. There may be some pragmatic infelicities in a theory of explanation that there are true explanations that we can't understand (fully or otherwise), but they are only that: pragmatic infelicities. They may strike the ear awkwardly. But they are not principled shortcomings.

The story of science, of course, is not complete. But there is also no going back. We now know that a single good theory can deliver millions of people from reasoning in darkness.

We saw it in the case of Newtonian science. The same promise exists for science-based policy. The urge to understand, prompted by wonder and quenched by explanation, ends in service to human well-being. But there a new cycle begins. After all, while mature science is acting as an instrument to inform policy, it can also uncover new wonders of human well-being.

NOTES

Chapter 1

1. Keltner and Haidt (2003) argue that awe results from two experiences: vastness and the need for accommodation.
2. One night, the 19th-century chemist August Kekule labored over the structure of compounds until he "fell into a reverie, and lo, the atoms were gamboling before my eyes. Whenever, hitherto, these diminutive beings had appeared to me, they had always been in motion; but up to that time, I had never been able to discern the nature of their motion. Now, however, I saw how, frequently, two smaller atoms united to form a pair; how a larger one embraced two smaller ones; how still larger ones kept hold of three or even four of the smaller; whilst the whole kept whirling in a giddy dance. I saw how the larger ones formed a chain. . . . I spent part of the night putting on paper at least sketches of these dream forms" (Rothenberg, 1993, 291).
3. This sense of understanding is most likely at the bottom of many standards for good explanation, such as simplicity (consider Watson's claim about his and Crick's completed DNA model that "a structure this pretty just had to exist" [1968/1999, 205]) or the rendering of the unfamiliar to the familiar. These standards, however, are complicated. What counts as simple or familiar is theory dependent, and not surprisingly, verdicts in particular cases are controversial. For example, the familiarity

account may seem to be violated by quantum mechanics, in which it appears that the familiar is being explained in terms of the less familiar. But this appearance may be misleading. An idea that "feels unfamiliar" to the outsider may convey a feeling to the scientist that the pieces of a theoretical puzzle have just fallen into place. Only a comprehensive overview can address all of these issues. The reader is referred to Salmon (1992, 14), where the reduction of the unfamiliar to the familiar is explicitly discussed.

4. Most pragmatists would apply this standard, as would constructive empiricists like van Fraassen, who gives explanation a merely pragmatic rather than epistemic status.
5. Petrucci and Harwood, 1997, 5.
6. Pope, ca. 1730/1954.
7. Nisbett and Ross, 1980.

Chapter 2

1. For this simple and approachable account of explanation, see Miller, 1987.
2. Trout, 2002, 2007.
3. Humphreys, 1989, 103.
4. See McAllister (1996) for a lovely philosophical discussion of these issues as they occur in science.
5. Watson, 1968/1999, 205.
6. Weinberg, 1994, 25.
7. Perkins, Allen, and Hafner, 1983.
8. Haidt, 2001, 821.
9. See Plantinga, 1993, 91–92.
10. Gombrich, 1984; also see Maritain, 1966; also see Solso, 1997.
11. Checkosky and Whitlock, 1973.
12. Reber, Winkielman, and Schwarz, 1998.
13. Tenney, Spellman, and MacCoun, 2008; Sah, Moore, and MacCoun, 2013.
14. Tetlock, 2005.
15. Penrod and Cutler, 1995.
16. Van Swol and Sniezek, 2005; Zarnoth and Sniezek, 1997.
17. Price and Stone, 2004.

18. Loewenstein, 1994.
19. Bradfield and Wells, 2000; Price and Stone, 2004; Thomas and McFadyen, 1995; Yates, Price, Lee, and Ramirez, 1996; and Zarnoth and Sniezek, 1997.
20. Cho and Schwarz, 2006.
21. Reber and Schwarz, 1999.
22. McGlone and Tofighbakhsh, 2000.
23. Because we use fluency, with its characteristic feeling or phenomenology, as a cue to truth, it is natural to ask how reliable a cue it is. Reber and Unkelbach (2010) look at the reliability of this fluency cue to judge a statement's truth. What would count as reliable? They propose a brute standard of justification for the reliance on this cue: "[A]s long as the percentage of true factual statements that are fluently processed is above 50%, the use of fluency is justified" (579). Most of that good fortune is ecological, since we fluently process most of the practical truths in our environment. In some contexts, this standard will hold up, where the truths are equisignificant and the run is long. This is unlikely to be the case in science, in which we are constantly testing ever more specific truths and ramping up the severity of the tests.
24. Kuhn, 1962/1996, 20.
25. Kuhn, 1962/1996, 25.
26. Kuhn, 1962/1996, 123.
27. Kuhn, 1962/1996, 192.
28. Thagard, 1989.
29. The feeling of understanding ranges from the grand, oceanic feeling that Freud talked about in religious experience to the tiny jolt you get from solving an anagram. Most explanations are more like the latter, what Kuhn called puzzle solving. The vast majority of scientific research is piecing together the ever-expanding menu of laboratory findings.
30. Copernicus, 1543/1952, 514.
31. James, 1890/1981, Chapter 26, 1136. Early on, Kruglanski (1989) postulated motivations for lay epistemic activities—many of them explanatory, and central to constructs he discusses, like the need for closure. Some of these motivations share features of the sense of understanding.
32. Paracelsus, 1665, Chapter VII.

33. Hertwig, Herzog, Schooler, and Reimer, 2008. After seeing the blustery endorsement of feelings of fluency, it would be disappointing in the extreme to find that it is generally an unreliable guide to the truth; it is a poor indicator of actual frequencies in the world. And where it is accurate, its success is, some portion of the time, a fragile trick, good for only one kind of carefully orchestrated quest. But this is what we find. When psychologists have turned to examine the "validity" of feelings of fluency, the rare successes don't form a natural grouping. It is as though psychologists spilled out all possible mental tasks onto our palette and picked them over until they found a few that managed to survive experimental tests. This would explain the odd assortment of cognitive jobs on the experimenters' list. The most expansive study looks at fluency feelings when charged with recognizing cities, companies, musical artists, athletes, and billionaires. What do these objects have in common? Nothing really. But to transform the feeling of fluency into an accurate indicator of true belief, they must be presented to people in a very controlled way.
34. Aristotle, 1970.
35. Aristotle, 1970.
36. Descartes, 1637/1965.
37. Weinberg, 2001, 24–25.
38. Feynman, 1965, 171.
39. Posner and Keele, 1968.
40. Rhodes and Tremewan, 1996; also see Rhodes, Yoshikawa, Clark, Lee, McKay, and Akamatsu, 2001.
41. Halberstadt and Rhodes, 2000, 2003.
42. Martindale and Moore, 1988.
43. Whitfield and Slatter, 1979.
44. Halberstadt and Rhodes, 2000, 2003.
45. Armstrong, Gleitman, and Gleitman, 1983.
46. Posner and Keele, 1968; also see Rosch and Mervis, 1975, and Mervis and Rosch, 1981.
47. Weisberg, Keil, Goodstein, Rawson, and Gray, 2008. For an analysis, see Trout, 2008.
48. Winkielman, Schwarz, Fazendeiro, and Reber, 2003.
49. Weisberg, Keil, Goodstein, Rawson, and Gray, 2008, 471.

50. Kruger and Dunning, 1999.
51. Keil, 2006.
52. Shah and Oppenheimer, 2007.
53. See Whittlesea and Leboe (2003) for a summary of the evidence.
54. Kuhn, 1962/1996, 162.
55. Alligators and turtles do have a temperature-sensitive period in the egg during which sex is determined.
56. Strictly speaking, brachycephaly also describes a developmentally normal type of skull with a high cephalic index.
57. Alter, Oppenheimer, Epley, and Eyre, 2007.
58. Soussignan, 2002.
59. In fact, for present purposes, it doesn't even matter how we categorize all of these processes that lead to speed and ease of processing "fluency." A completed psychological theory may taxonomize these effects in a way that is quite different from the one I use here. The important feature is that they transform specific dimensions into easier processing that carries a phenomenological sense of understanding or accuracy.

Chapter 3

1. Kounios and Beeman, 2009.
2. Kounios and Beeman, 2009, 210.
3. Mednick, 1962.
4. Kounios, Frymiare, Bowden, Fleck, Subramaniam, Parrish, and Jung-Beeman, 2006.
5. Woolf, 1979, 140.
6. Hobbes, 1651/1994, Book 1, Chapter 6, paragraph 7.
7. Brembs, Lorenzetti, Reyes, Baxter, and Byrne, 2002.
8. Skurnik, Yoon, Park, and Schwarz, 2005.
9. Ruskin, 1887, 124.
10. Johnson, 1840, 163.
11. Litman, 2005, 793; also see Berridge and Kringelbach, 2015.
12. Litman, 2005.
13. Of course, it might be that formal features of an explanation also drive the acceptance of an inference to best explanation, in addition to the sense of understanding it supplies.

In some quarters, for example, philosophers think that formal features of a good explanation can be used to produce a prediction. On this view, a good explanation has logical or temporal/tensed features whose direction can be transformed with illuminating results. But first, it is not clear how strong a constraint this places on good explanation, and second, in practice, these transformations of formal features, found in uses of retrodiction, for example, seem strained or unnatural for subject matters like evolutionary theory or historical geology, whose function is not really to predict. After all, the temporal component in retrodiction concerns what you will find if you, say, choose to look in a particular stratum for an index fossil. It does not concern what the theory says a particular species will evolve into. The latter is the natural sense of "prediction" found in the physical sciences, the former a "retrodiction," a kind of post hoc prediction and formal feature inaugurated by philosophers of science to preserve a role for prediction in theories that don't have much of a need for it but perhaps require it for official scientific status.

Chapter 4

1. Railton, 1981, 242.
2. Reichenbach, 1938/2006.
3. Boyd, 1983, 195.
4. Lipton, 1991, 58.
5. See Hempel, 1945.
6. Glymour, 1985, 99.
7. Fine, 1986, 161.
8. Note, however, that the experimental method works *equally* poorly with bad theories. This is not a problem merely for the realist, but for those that champion the standard view of scientific progress.
9. Gilbert Harman's 1965 paper "Inference to the Best Explanation" is widely cited, but its message—that even enumerative induction is based on explanatory principles—is routinely ignored. I plan to resurrect and fortify this position.

Evidence of renewed interest in "explanationism" and abductive reasoning is found in the new generation of scholarship in the psychology of explanation. See Lombrozo (2012) for one such example.

10. Strevens, 2013b.
11. This way of explaining the success of inductive reasoning may appear to commit me to an internalist epistemology. This would be troubling, because I have committed elsewhere (Bishop and Trout, 2005) to a version of reliabilism. But talk of explanatory reasoning does not entail internalism. In fact, because my account of explanation is ontic, the reliability of explanatory reasoning may result from being appropriately related to the causal mechanisms and processes that make the explanation true, and not necessarily the ones identifiable by the standard internalist cues.
12. Quine, 1970, 50.
13. See, for example, Finger, 2001.
14. Boyd, 1983.
15. Pyle, 1995.
16. Stanford, 2006, 160, footnote 3.
17. See the National Institute of Standards and Technology (2015) for descriptive illustrations of the process.
18. Lipton, 2000, 184.
19. van Fraassen, 1989, 132.
20. Laudan, 1981.
21. Fine, 1986, 161.
22. Fine, 1991, 82.
23. Stanford, 2006, 145, emphasis in original.
24. Lipton, 2001, 351; http://www.hps.cam.ac.uk/people/lipton/quests_of_a_realist.pdf.
25. See, for instance, Laudan (1981) and Fine (1986).
26. Hardin and Rosenberg, 1982, 610.
27. Stanford, 2006, 160, footnote 3.
28. Glymour, 1985, 116.
29. Boyd, 1983, 217.
30. This is the idea behind population-guided estimation in Trout (1998).

Chapter 5

1. von Euxkull, 1934/1957. For another use of the notion of an Umwelt, see Clark (1997).
2. Lutz, 2005.
3. Arnason, Hart, and O'Connell-Rodwell, 2002.
4. Bromm, Hensel, and Tagmat, 1976.
5. Bingman, 1998.
6. Quoted in Angier, 2008.
7. Quoted in Angier, 2008.
8. Quoted in Angier, 2008.
9. Carruthers, 2008.
10. Diamond, 1999.
11. See Bacon, 1620/2000.
12. Kuhn, 1962/1996, 19–20.
13. Kuhn, 1962/1996, 31–32.
14. Silverberg, 2000, 12.
15. Minkoff and Baker, 2001, 6.
16. Starr and Taggart, 1984, 21.
17. See Nummendal (2007, 1–2) for her account of later 16th-century alchemist Philipp Sömmering.
18. For one example in which contingency plays a role, see Kitcher (1990).
19. Priestley, 1775, 114; Pages numbers found at http://books.google.com/books?id=gB0UAAAAQAAJ&printsec=frontcover#v=onepage&q&f=false.
20. Priestley, 1775, 29.
21. Pasteur, 1854.
22. From his presidential address (November 24, 1877) to the Philosophical Society of Washington. Reprinted in Bauer (1908).
23. Fodor, 1974, 202.
24. Kuhn, 1962/1996, 152–153.
25. Rogers, 2011, 65.
26. This form of argument is widely called path dependent.
27. Galileo tells this story in more detail in his "The Assayer" (1623/1957).

28. I am not arguing that the existence of any specific religion is a contingent fact. Instead, I am making the point that, in regard to the science of timekeeping, religion is a contingent factor. That is, certain scientific discoveries may have depended on the existence and development of a particular religious or cultural view.
29. For an alternative, but still pleasingly ontic account, see Craver (2007), especially p. 200: "I advocate an ontic view of explanation according to which one explains a phenomenon by showing how it is situated in the causal structure of the world." A brief philosophical digression. It is reasonable to wonder whether the view I have proposed is really a theory of explanation. If the ontic view is one about the very nature of explanation, then it is an open question how true a schema has to be in order to be an explanation. On the other hand, if it's a view about what makes an explanation good, then one might object that I haven't really provided an account of what explanations are. I prefer the latter. I don't think I have offered a theory of explanation, but I am not sure that counts as an admission, or if so whether it is damaging. I think we can say at most only a few things about common practices associated with what we tend to call "explanations." It is clear from Chapter 2 that I think explanation is an important but informal affair. I doubt we can lay down many general rules that are true of all and only explanations.
30. Strevens, 2013a.
31. Cowan, 2005.
32. Just and Carpenter, 1992.
33. Halford, Wilson, and Phillips, 1998.
34. Oberauer and Kliegl, 2001.
35. Reber and Unkelbach, 2010.
36. Skurnik, Yoon, Park, and Schwarz, 2005.
37. Lynch and Medin, 2006.
38. University of California, San Francisco Medical Center, 2015.
39. Scicurious [Brookshire, B.], 2013.
40. Carrington, 2013.
41. Bravender, 2009.

42. For this latter question, see Talhelm, Zhang, Oishi, Shimin, Duan, Lan, and Kitayama (2014).
43. Stevens, 2008, 176.
44. Haldane, 1914, 58.
45. Haldane, 1914, 55.
46. Haldane, 1914, 58.
47. Haldane, 1927.
48. Cartwright, 1983, 101f.
49. A conceptual treatment of this issue is found in McGinn (1991).
50. Kuhn, 1962/1996, 118.
51. Kuhn, 1962/1996, 116–117.
52. Kuhn, 1962/1996, 120.
53. Trout, 1994.

Chapter 6

1. A beautiful and careful treatment of the issues separating and drawing together Sennert and Boyle can be found in Newman (2006). Chalmers (2009) also explores corpuscular themes at the border of philosophy and science.
2. Perrin, 1913, 215–216.
3. Perrin, 1913, 215–216.
4. Dobbs, 1982.
5. Newman, 2006, 166–167.
6. Kuhn, 1962/1996, 56.
7. Galen, 1916.
8. Park, 1990.
9. Crombie, 1953.
10. This graph represents an unselective catalog of events cited as noteworthy in a comprehensive and current history of science by McClellan and Dorn (1999). The goal was to represent events that contributed to extending human control over the world or our understanding of nature. We recorded any event the authors deemed worth mentioning. Our unselective standard inclined us to equally weight achievements that were technological, mathematical/formal, merely descriptive, and theoretical.

It is likely that scientific progress will be confounded partly, but not entirely, with the rise of tools for disseminating science. For example, the invention of the printing press made it easier to create and distribute manuscripts, and the rise of scientific societies created coordinated networks of collaboration and intellectual exchange, making it easier to disseminate scientific findings.

This unselective sample of achievements represents exactly the pace and sequence conveyed in anecdotal descriptions of the history of science: an overall upward trend, with no recessions and a steep ascent for theoretical discovery in the modern and postindustrial period. But notice, the shape of this curve is not inevitable. I want to thank Spencer Weart for practical suggestions about performing this count, and for advice about proxies for progress.

11. Geoffrey, 1718.
12. Williams, 2010.
13. Watson, 1968/1999, 163.
14. Wilson, 2008, 291.
15. Quoted in Koyré, 1965, 15.
16. Quoted in Gillispie, 1997, 204.
17. Dobbs, 1982, 513.
18. Just 30 years before, in 1571, Philipp Sömmering (the court alchemist of Duke Julius of Braunschweig-Wolfenbüttel) and his assistants were ultimately charged with fraud and with a variety of occult actions like sorcery—not for practicing alchemy, but for doing it badly for a love potion and too well for an invisibility formula (Nummedal, 2007).
19. Newton, 1730/1952, 375–376.
20. Newton, 1730/1952, 376.
21. Newton, 1730/1952, 394–395.
22. This is how one distinguished history of chemistry describes the mystical symbolism of alchemy:

    > The imagery of Christian alchemy is well known. Saint George slaying the dragon, a king and queen entering a bath under threat of a naked sword brandished by a soldier, a wolf devouring a dead king, the fall of Troy and the

death of Priam, aged Mercury simmering in a bath until his spirit, a white dove, escapes: each of these images is crowded with theological mystic, mythic, astrological, and operational references (Morazé, 1986). The soldier was the solvent who would force the sulfur king and mercury queen to react with each other; the death of Priam was the dissolution or melting of a compound; the hermetic dragon, who represented a test upon the alchemist's path, a counterforce opposing his work, guarded the grotto where Quintessence was found; the wolf was the antinomy devouring the gold before the purifying fire regenerated a living, active, "philosopher" king; and so on (Bensaude-Vincent and Stengers, 1996, 18).

23. But it may also take several smaller hunches to culminate in a major advance, to reach that threshold. In mechanics, it was reached around the time of Boyle's treatment of constituents as corpuscles, and Newton assumed as much in his laws of motion. But there wasn't a single Newtonian hunch, even in Newton's field.
24. Let us take (1) first. Roundness, hardness, and elastic collisions are important features of objects, not incidental window dressing. They explain a wide range of the behavior of systems that contain them. Heat and pressure increase with the "agitation" or motion of particles. In general, their momenta get transferred fairly efficiently to the corpuscles they hit.

    Second, while there were many alternatives to the corpuscular view that might have individually explained the behavior of heated, enclosed containers, the buoyancy of an airplane wing, or the striking changes that occur in a liquid during a titration or precipitation, the corpuscular approach handles all at once.

    Third, in retrospect, quantitative investigations do seem to vindicate not just the existence but the character of corpuscles. There were laws about their interaction. They turn up unexpectedly in other domains, like thermodynamics and pneumatic chemistry. So the corpuscular hypothesis is a fairly accurate description of an underlying reality.

25. Newton, 1730/1952, 389.
26. Meinel, 1988, 78.
27. Meinel, 1988, 93.
28. Newman, 2006, 44.
29. These powers vary in degree. Sennert explained the heat and boiling when dissolving a substance in acid as "the sudden motion of similar to similar." Newton explained this chain of events in exactly the same way: the torrential heat produced by these reactions resulted from the imperceptibly small motion of the atoms racing toward one another. The image behind this passage is vivid. In effect, the corpuscles of the dissolving substance display such an attraction for the corpuscles of aqua fortis that they rush toward them with a velocity that causes heat and boiling. Once bonded, you can make them bond with another substance, or they can precipitate. For a more thorough discussion of this issue, see Newman (2006, 134–135).
30. In the world of late medieval and renaissance Europe, sanitation was a challenge. Sewage was everywhere, from the street to your front door. It might have come from an enemy or just a lazy neighbor (Carlin, 1996). Either way, van Helmont wanted none of it:

    > Hath any one with his excrements defiled the threshold of thy door, and thou intendest to prohibit that nastiness in the future, do but lay a red-hot iron upon the excrement, and the immodest sloven shall, in a very short space, grow scabby on his buttocks, the fire torrifying the excrement, and by dorsal magnetism driving the antinomy of the burning, into his impudent anus (Van Helmont, 1650, 13).

    This somewhat mystical appeal to backward causation is common in the later alchemy that gave rise to a consistent corpuscularism. As one modern commentator put it:

    > Consider Kenelm Digby's 1658 account of the weapon salve, or the treatment of wounds at a distance by

manipulation of the weapon that caused them. Digby in fact offered a fascinating, sophisticated application of early modern corpuscularianism, yet many philosophers today suppose that to take an interest in a false theory from the past such as this one, to research it and to write about it, implies a rejection of the idea of truth itself (Smith, 2011).

Paracelsus wrote:

> Take of moss growing on the head of a thief who has been hanged and left in the air; of real mummy; of human blood, still warm—of each one ounce; of human suet, two ounces; of linseed oil, turpentine, and Armenian bole—of each two drachms. Mix all well in a mortar, and keep the salve in an oblong, narrow urn (Goclenius, 1613, 95).

31. Debus, 1967.
32. Dobbs, 1975; also see Newman, 2006.
33. This puzzle is first set out in Boyd, 1980, 636.
34. As a mark of its contingency, Newton's mature work explores alchemical themes on the nature of light. He concludes that blackness comes from putrefaction, a move that leads one respected Newton scholar to say that "the whole of his career after 1675 may be seen as one long attempt to integrate alchemy and the mechanical philosophy (Dobbs, 1975, 230).
35. Bosveld, 2010.
36. Perhaps Newton had learned well from the case of Copernicus, rescued from religious persecution by Ossiander's conditional, merely formal or mathematical gloss on heliocentrism: "Whether elastic fluids do really consist of particles so repelling each other, is a physical question. We have here demonstrated mathematically the property of fluids consisting of particles of this kind, that hence philosophers may take occasion to discuss that question" (quoted in Brush, 2003, 54).

    So there was a huge theoretical gap between what Newton believed and what he officially concluded. This is not the toothless logical point that all hypotheses are underdetermined by the evidence, which is true at all times but normally inoffensive. Even given that, he still made a significant guess, and for historically contingent reasons, the hunch proved accurate, and the guess paid off.

37. For a very nice overview, see Turner (1995).
38. Quoted in Todes, 1997, 228.
39. Quoted in Todes, 1997, 211.
40. Ferris, 2010, 4.
41. van Fraassen, 1980, 40.

Chapter 7

1. See Johnson (2014).
2. See coverage of A.K. Pradeep's media coverage in his role as NeuroFocus chairman: http://www.nytimes.com/2010/11/14/business/14stream.html?_r=0.
3. See Trout (2009) for an exploration of this theme.
4. One of the best collections containing psychologists examining explanation is Keil and Wilson (2000). Also see Holyoak and Morrison (2012).
5. Rozenblit and Keil, 2002; Mills and Keil, 2004.
6. Rozenblit and Keil, 2002.
7. Fernbach, Rogers, Fox, and Sloman, 2013.
8. Bishop (2015) marshals much of this empirical evidence.

REFERENCES

Angier, N. 2008. Scientists and philosophers find that "gene" has a multitude of meanings. *New York Times*, November 10. Retrieved from http://www.nytimes.com/2008/11/11/science/11angi.html?pagewanted=all&_r=0.

Alter, A.L., D.M. Oppenheimer, N. Epley, and R.N. Eyre. 2007. Overcoming intuition: Metacognitive difficulty activates analytic reasoning. *Journal of Experimental Psychology* 136: 569–576.

Aristotle. 1970. *The history of animals*. Trans. A.L. Peck. Cambridge, MA: Harvard University Press.

Armstrong, S.L., L. Gleitman, and H. Gleitman. 1983. What some concepts might not be. *Cognition* 13: 263–308.

Arnason, B.T., L.A. Hart, and C.E. O'Connell-Rodwell. 2002. The properties of geophysical fields and their effects on elephants and other animals. *Journal of Comparative Psychology* 116, no. 2: 123–132.

Bacon, F. 1620/2000. *The new organon*. Ed. L. Jardine and M. Silverthorne. Cambridge: Cambridge University Press.

Bauer, L.A. 1908. The instruments and methods of research. *Philosophical Society of Washington Bulletin* 15: 103–126.

Bensaude-Vincent, B., and I. Stengers. 1996. *A history of chemistry*. Cambridge, MA: Harvard University Press.

Berridge, K., and Kringelbach, M. 2015. Pleasure systems in the brain. *Neuron*. 86(3): 646–164.

Bingman, V.P. 1998. Spatial representations and homing pigeon navigation. In S. Healy (ed.), *Spatial representation in animals*. Oxford: Oxford University Press, pp. 69–85.

Bishop, M. 2015. *The good life: Unifying the philosophy and psychology of well-being*. New York: Oxford University Press.

Bishop, M. and J.D. Trout. 2005. *Epistemology and the psychology of human judgment*. New York: Oxford University Press.

Bosveld, J. 2010. Isaac Newton, world's most famous alchemist. *Discover Magazine*, July–August. Retrieved from http://discovermagazine.com/2010/jul-aug/05-isaac-newton-worlds-most-famous-alchemist#.UScHnaVweSq.

Boyd, R. 1980. Scientific realism and naturalistic epistemology. *Proceedings of the Biennial Meeting of the Philosophy of Science Association* 2: 613–666.

———. 1983. On the current status of scientific realism. In R. Boyd, P. Gasper, and J.D. Trout (eds.), *Philosophy of science*. Boston: MIT Press, pp. 195–222.

Bradfield, A., and G. Wells. 2000. The perceived validity of eyewitness identification testimony: A test of the Biggers criteria. *Law and Human Behavior* 24: 581–594.

Bravender, R. 2009. Study links smog exposure to premature death. *The New York Times*, March 12. Retrieved from http://www.nytimes.com/gwire/2009/03/12/12greenwire-study-links-smog-exposure-to-premature-death-10098.html.

Brembs, B., F. Lorenzetti, F. Reyes, D. Baxter, and J. Byrne. 2002. Operant reward learning in aplysia: neuronal correlates and mechanisms. *Science* 296: 1706–1708.

Bromm, B., H. Hensel, and A.T. Tagmat. 1976. The electrosensitivity of the isolated ampulla of lorenzini in the dogfish. *Journal of Comparative Psychology* 111, no. 2: 127–136.

Brush, S. 2003. *The kinetic theory of gases*. London: Imperial College Press.

Carlin, Martha. 1996. *Medieval Southwark*. London: Hambledon Press.

Carrington, D. 2013. Whales flee from military sonar leading to mass strandings, research shows. *The Guardian*, July

3. Retrieved from http://www.theguardian.com/environment/2013/jul/03/whales-flee-military-sonar-strandings.

Cartwright, N. 1983. *How the laws of physics lie*. Oxford: Oxford University Press.

Carruthers, P. 2008. Meta-cognition in animals: A skeptical look. *Mind and Language* 23, no. 1: 58–89.

Chalmers, A. 2009. *The scientist's atom and the philosopher's stone*. New York: Springer.

Checkosky, S.F., and D. Whitlock. 1973. Effects of pattern goodness on recognition time in a memory search task. *Journal of Experimental Psychology* 100: 341–348.

Cho, H., and N. Schwarz. 2006. If I don't understand it, it must be new: Processing fluency and perceived product innovativeness. *Advances in Consumer Research* 33: 319–320.

Clark, A. 1997. *Being there: putting brain, body, and world together again*. Cambridge, MA: MIT Press/Bradford Books.

Copernicus, N. 1543/1952. *Revolutions of heavenly spheres*, Book I, Section 4. In M. Adler (ed.), *Great books of the Western world*, vol. 16. Trans. C. Glenn Wallis. Chicago: William Benton Publishers.

Cowan, N. 2005. *Working memory capacity*. Hove, East Sussex: Psychology Press.

Craver, C. 2007. *Explaining the brain*. New York: Oxford University Press.

Crombie, A.C. 1953. *Robert Grossteste and the origins of experimental science, 1100-1700*. Oxford: Clarendon Press.

Debus, A. 1967. Fire analysis and the elements in the sixteenth and the seventeenth centuries. *Annals of Science* 23: 127–147.

Descartes, R. 1637/1965. *Discourse on method, optics, geometry, and meteorology*. Trans. P.J. Olscamp. Indianapolis: Bobbs-Merrill.

Diamond, J. M. 1999. *Guns, germs, and steel: The fates of human societies*. New York: W.W. Norton & Company.

Dobbs, B.J.T. 1975. *The foundations of Newton's alchemy*. Cambridge: Cambridge University Press.

———. 1982. Newton's alchemy and his theory of matter. *Isis* 73: 511–528.

Fernbach, P., T. Rogers, C. Fox, and S. Sloman. 2013. Political extremism is supported by an illusion of understanding. *Psychological Science* 24, no. 6: 939–945.

Ferris, T. 2010. *The science of liberty*. New York: Harper Collins.

Feynman, R.P. 1965. *The character of physical law*. Cambridge, MA: MIT Press.

Fine, A. 1986. Unnatural attitudes: Realist and instrumentalist attachments to science. *Mind* 95: 149–179.

———. 1991. Piecemeal realism. *Philosophical Studies* 61, no. 1–2: 79–96.

Finger, S. 2001. *Origins of neuroscience: A history of explorations into brain function*. New York: Oxford University Press.

Fodor, J. 1974. *The language of thought*. New York: Crowell.

Geoffrey, E. 1718. Table of affinities. *Memoir to the Academy of Sciences*, 202–212. (For an English translation, see *Science in Context*, 9, 313–320 (1996).

Galen. 1916. *On the natural faculties*. Trans. A.J. Brock. London: William Heinemann.

Galileo. 1623/1957. The Assayer. In *Discoveries and opinions of Galileo*. Trans. S. Drake. New York: Doubleday Anchor.

Gillispie, G.C. 1997. *Pierre-Simon Laplace: A life in exact science*. Princeton, NJ: Princeton University Press, p. 204.

Glymour, C. 1985. Explanation and realism. In P. Churchland (ed.), *Images of science*. Chicago: University of Chicago Press.

Goclenius, R. 1613. *Tractatus de Magnetica Vulnerum Curatione*. Francofurti: imp. P. Musculi et R. Pistorii.

Gombrich, E. H. 1984. *The sense of order: A study in the psychology of decorative art*. London: Phaidon Press.

Haidt, J. 2001. The emotional dog and its rational tail: A social intuitionist approach to moral judgment. *Psychological Review* 108: 814–834.

Halberstadt, J., and G. Rhodes. 2000. The attractiveness of non-face averages: Implications for an evolutionary explanation of the attractiveness of average faces. *Psychological Science* 11: 285–289.

———. 2003. It's not just average faces that are attractive: Computer-manipulated averageness makes birds, fish,and automobiles attractive. *Psychonomic Bulletin & Review* 10: 149–156.

Haldane, J.S. 1914. *Mechanism, life, and personality: An examination of the mechanistic theory of life and mind*. London: J. Murray.

Haldane, J.B.S. 1927/2001. Possible worlds. Piscataway, NJ: Transaction Publishers, Reprint edition.

Halford, G., W. Wilson, and S. Phillips. 1998. Processing capacity defined by relational complexity: Implications for comparative, developmental, and cognitive psychology. *Behavioral and Brain Sciences* 21: 803–864.

Harman, G. 1965. The inference to the best explanation. *Philosophical Review* 74, no 1: 88–95.

Hardin, C., and A. Rosenberg. 1982. In defense of convergent realism. *Philosophy of Science* 49, no. 4: 604–615.

Hellemans, A., and B. Bunch. 1988. *The timetables of science: a chronology of the most important people and events in the history of science.* New York: Simon and Schuster.

Hempel, C. 1945. Studies in the logic of confirmation. *Mind* 54: 1–26.

Hertwig, R., S. Herzog., L. Schooler, and T. Reimer. 2008. Fluency heuristic: A model of how the mind exploits a by-product of information retrieval. *Journal of Experimental Psychology: Learning, Memory, and Cognition* 34: 1191–1206.

Hobbes, T. 1651/1994. *Leviathan, with selected variants from the Latin edition of 1668.* Ed. E. Curley. Indianapolis: Hackett.

Holyoak, K., and R. Morrison (eds.). 2012. *Oxford handbook of thinking and reasoning.* Oxford: Oxford University Press, pp. 260–276.

Humphreys, P. 1989. *The chances of explanation.* New York: Oxford University Press.

James, W. 1890/1981. *Principles of psychology,* vol 2. Cambridge, MA: Harvard University Press.

Johnson, S. 1840. *The works of Samuel Johnson,* vol. 1. New York: Alexander and Blake.

Johnson, B. 2014. How one GOP-controlled committee is waging war on science. *Huffington Post,* June 24. Retrieved from http://www.huffingtonpost.com/2014/06/24/house-science-committee_n_5525609.html?ir=Education.

Just, M., and P. Carpenter. 1992. A capacity theory of comprehension: Individual differences in working memory. *Psychological Review* 99, no. 1: 122–149.

Keil, F., and R. Wilson. 2000. *Explanation and cognition*. Cambridge, MA: MIT Press/Bradford Books.

Keil, F.C. 2006. Explanation and understanding. *Annual Review of Psychology* 57: 227–254.

Keltner, D., and J. Haidt. 2003. Approaching awe, a moral, spiritual, and aesthetic emotion. *Cognition & Emotion*, 17(2), 297–314.

Kitcher, Philip. 1990. The division of cognitive labor. *Journal of Philosophy* 87: 5–22.

Koyré, A. 1965. *Newtonian studies*. Cambridge, MA: Harvard University Press.

Kounios, J., J. Frymiare, J. Bowden, J. Fleck, K. Subramaniam, T. Parrish, and M. Jung-Beeman. 2006. The prepared mind: Neural activity prior to problem presentation predicts subsequent solution by sudden insight. *Psychological Science* 17, no. 10: 882–890.

Kounios, J., and M. Jung-Beeman. 2009. The aha! moment: The cognitive neuroscience of insight psychological science. *Current Directions in Psychological Science* 18, no. 4: 210–216.

Kruger, J., and D. Dunning. 1999. Unskilled and unaware of it: How difficulties in recognizing one's own incompetence lead to inflated self-assessments. *Journal of Personality and Social Psychology* 77: 1121–1134.

Kruglanski, A. 1989. *Lay epistemics and human knowledge*. New York: Plenum.

Kuhn, T.S. 1962/1996. *The structure of scientific revolutions*, 3rd ed. Chicago: University of Chicago Press.

Laudan, L. 1981. A confutation of convergent realism. *Philosophy of Science* 48, no. 1: 19–49.

Litman, J. 2005. Curiosity and the pleasures of learning: Wanting and liking new information. *Cognition and Emotion* 19, no. 6: 793–814.

Lipton, P. 1991. *Inference to the best explanation*. London: Routledge.

———. 2000. Inference to the best explanation. In W.H. Newton-Smith (ed.), *A companion to the philosophy of science*. Oxford: Blackwell, pp. 184–193.

———. 2001. Quests of a realist. *Metascience* 10, no. 3: 347–353.

Loewenstein, G. 1994. The psychology of curiosity: A review and reinterpretation. *Pscyhological Bulletin* 116: 75–98.

Lombrozo, T. 2012. Explanation and abductive inference. In K. Holyoak and R. Morrison (eds.), *Oxford handbook of thinking and reasoning*. Oxford: Oxford University Press, pp. 260–276.

Lutz, D. 2005. *Tuatara: A living fossil*. Salem, OR: Dimi Press.

Lynch, E., and D.L. Medin. 2006. Explanatory models of illness: A study of within-culture variation. *Cognitive Psychology* 53, no. 4: 285–309.

Maritain, J. 1966. *Creative intuition in art and poetry*. Paris: Desclée de Brouwer.

Martindale, C., and K. Moore. 1988. Priming, prototypicality, and preference. *Journal of Experimental Psychology: Human Perception and Performance* 14: 661–670.

McAllister, J. 1996. *Beauty and revolution in science*. Ithaca, NY: Cornell University Press.

McClellan, J., and H. Dorn. 1999. *Science and technology in world history: An introduction*. Baltimore: John Hopkins University Press.

McGinn, C. 1991. *The problem of consciousness*. London: Blackwell.

McGlone, M.S., and J. Tofighbakhsh. 2000. Birds of a feather flock conjointly (?): Rhyme as reason in aphorisms. *Psychological Science* 11: 424–428.

Mednick, S. 1962. The associative of the creative process. *Psychological Review* 69, no. 3: 220–232.

Meinel, C. 1988. Early seventeenth-century atomism: Theory, epistemology, and the insufficiency of experiment. *Isis* 79: 68–103.

Mervis, C., and Rosch, E. 1981. Categorization of natural objects. *Annual Review of Psychology* 32: 89–115.

Miller, R. 1987. *Fact and method*. Princeton, NJ: Princeton University Press.

Mills, C., and F. Keil. 2004. Knowing the limits of one's understanding: The development of an awareness of an illusion of explanatory depth. *Journal of Experimental Child Psychology* 87: 1–32.

Minkoff, E.C., and P.J. Baker. 2001. *Biology today: An issues approach*. New York: Garland Publishing.

Morazé, C. 1986. *Les Origines Sacrées des Sciences Modernes*. Paris: Fayard.

National Institute of Standards and Technology. 2015. Time measurement and analysis service. Retrieved from http://www.nist.gov/pml/div688/grp40/tmas.cfm.

Newman, W.R. 2006. *Atoms and alchemy: Chymistry and the experimental origins of the scientific revolution*. Chicago: University of Chicago Press.

Newton, I. 1730/1952. *Opticks*. New York: Dover Publications.

Nisbett, R.E., and L. Ross. 1980. *Human inference: Strategies and shortcomings of social judgment*. Englewood Cliffs, NJ: Prentice-Hall.

Nummedal, T. 2007. *Alchemy and authority in the Holy Roman Empire*. Chicago: University of Chicago Press.

Oberauer, K., and R. Kliegl. 2001. Beyond resources: Formal models of complexity effects and age differences in working memory. *European Journal of Cognitive Psychology* 13, no. 1–2: 187–215.

Paracelsus, T. 1665. Concerning the renovation of men. In *Of the tincture of the philosophers*. Trans. J.H. Oxon. London: J.H. Oxon, pp. 35–6.

Park, D. 1990. *The how and the why: An essay on the origins and development of Physical Theory*. Princeton, NJ: Princeton University Press.

Pasteur, L. 1854 (December 7). Lecture, University of Lille.

Penrod, S., and B. Cutler. 1995. Witness confidence and witness accuracy: Assessing their forensic relation. *Psychology, Public Policy, and Law* 1: 817–845.

Perkins, D., R. Allen, and J. Hafner. 1983. Difficulties in everyday reasoning. In W. Maxwell and J. Bruner (eds.), *Thinking: The expanding frontier*. Philadelphia: Franklin Institute Press, pp. 177–189.

Perrin, J. 1913. *Les Atomes*. Paris: Alcan.

Petrucci, R.H., and W.S Harwood. 1997. *General chemistry principles and modern applications*, 7th ed. Upper Saddle River, NJ: Prentice-Hall.

Plantinga, A. 1993. *Warrant and proper function*. New York: Oxford University Press.

Pope, A. 1954. Epigraph intended for Sir Isaac Newton. In N. Ault and J. Butt (eds.), *Poems of Alexander Pope, Vol. 6: Minor Poems*. London: Methuen.Ptolemy, p. 317.

Posner, M.I., and S.W. Keele. 1968. On the genesis of abstract ideas. *Journal of Experimental Psychology* 77: 353–363.

Priestley, J. 1775. *Experiments and observations on different kinds of air*, vol. II. London: J. Johnson.

Price, P., and E. Stone. 2004. Intuitive evaluation of likelihood judgment producers: Evidence for a confidence heuristic. *Journal of Behavioral Decision Making* 17, no. 1: 39–57.

Pyle, A. 1995. *Atomism and its critics: From Democritus to Newton.* South Bend, IN: St. Augustine's Press.

Quine, W.V. 1970. Natural kinds. In C. Hempel, D. Davidson, and N. Rescher (eds.), *Essays in honor of Carl G. Hempel.* Dordrecht: D. Reidel, pp. 41–56.

Railton, P. 1981. Probability, explanation, and information. *Synthese* 48: 233–256.

Reber, R., and N. Schwarz. 1999. Effects of perceptual fluency on judgments of truth. *Consciousness and Cognition* 8: 338–342.

Reber, R., P. Winkielman, and N. Schwarz. 1998. Effects of perceptual fluency on affective judgments. *Psychological Science* 9: 45–48.

Reber, R., and C. Unkelbach. 2010. The epistemic status of processing fluency as source for judgments of truth. *Review of Philosophical Psychology* 1: 563–581.

Reichenbach, H. 1938/2006. *Experience and prediction.* South Bend, IN: University of Notre Dame Press.

Rhodes, G., and T. Tremewan. 1996. Averageness, exaggeration and facial attractiveness. *Psychological Science* 7: 105–110.

Rhodes, G., S. Yoshikawa, A. Clark, K. Lee, R. McKay, and S. Akamatsu. 2001. Attractiveness of facial averageness and symmetry in non-western cultures: In search of biologically based standards of beauty. *Perception* 30: 611–625.

Rosch, E., and C.B. Mervis. 1975. Family resemblances: Studies in the internal structure of categories. *Cognitive Psychology* 7: 573–605.

Rogers, K. 2011. *Medicine and healers through history.* New York: Encyclopedia Britannica.

Rothenberg, A. 1993. Creative homospatial and janusian processes in kekule's discovery of the structure of the benzene molecule. In J.H. Wotiz (ed.), *The Kekule riddle.* Trans. F.R. Japp. Clearwater, FL: Cache River Press, pp. 286–309.

Rozenblit, L., and F. Keil. 2002. The misunderstood limits of folk science: An illusion of explanatory depth. *Cognitive Science* 92: 1–42.

Ruskin, J. 1887. *The two paths: Being lectures on art and its applications to decoration and manufacture.* Orpington, Kent: George Allen.

Sah, S., D. Moore, and R. MacCoun. 2013. Cheap talk and credibility: The consequences of confidence and accuracy on advisor credibility and persuasiveness. *Organizational Behavior and Human Decision Processes* 121: 246–255.

Salmon, W. 1992. Scientific explanation. In W. Salmon (ed.), *Introduction to the philosophy of science.* Englewood Cliffs, NJ: Prentice-Hall, pp. 7–41.

Scicurious [Brookshire, B.]. September 23, 2013 (September 23). IgNobels 2013! The dung beetle and the stars. *Scientific American.* Retrieved from http://blogs.scientificamerican.com/scicurious-brain/2013/09/23/ignobels-2013-the-dung-beetle-and-the-stars/.

Shah, A.K., and D.M. Oppenheimer. 2007. Easy does it: The role of fluency in cue weighting. *Judgment and Decision Making* 2: 371–379.

Silverberg, M.S. 2000. *Chemistry: The molecular nature of matter and change.* New York: McGraw-Hill.

Skurnik, I., C. Yoon, D. Park, and N. Schwarz. 2005. How warnings about false claims become recommendations. *Journal of Consumer Research* 31: 713–724.

Smith, J.E. 2011. The flight of curiosity. *New York Times*, May 22. Retrieved from http://opinionator.blogs.nytimes.com/2011/05/22/the-flight-of-curiosity/.

Solso, R.L. 1997. *Cognition and the visual arts.* Cambridge, MA: MIT Press.

Soussignan, R. 2002. Duchenne smile, emotional experience, and automatic reactivity: A test of the facial feedback hypothesis. *Emotion* 2, no. 1: 52–74.

Stanford, P.K. 2006. *Exceeding our grasp: Science, history, and the problem of unconceived alternatives.* New York: Oxford University Press.

Starr, C., and R. Taggart. 1984. *Biology: The unity and diversity of life,* 3rd ed. Belmont, CA: Wadsworth.

Stevens, M. 2008. *Depth: An account of scientific explanation.* Cambridge, MA: Harvard University Press.

Strevens, M. 2013a. No understanding without explanation. *Studies in History and Philosophy of Science* 44: 510–515.

Strevens, M. 2013b. *Tychomancy.* Cambridge, MA: Harvard University Press.

Talhelm, T., X. Zhang, S. Oishi, C. Shimin, D. Duan, X. Lan, and S. Kitayama. 2014 (May 9). Large-scale psychological differences within China explained by rice versus wheat agriculture. *Science* 344, no. 6184: 603–608.

Tenney, E.R., B.A. Spellman, and R.J. MacCoun. 2008. The benefits of knowing what you know (and what you don't): How calibration affects credibility. *Journal of Experimental Social Psychology* 44: 1368–1375.

Tetlock, P.E. 2005. *Expert political judgment: How good is it? How can we know?* Princeton, NJ: Princeton University Press.

Thagard, P. 1989. Explanatory coherence. *Behavioral and Brain Sciences* 12: 435–502.

Thomas, J., and R. McFadyen. 1995. The confidence heuristic: A game-theoretic analysis. *Journal of Economic Psychology* 16: 97–113.

Todes, D.T. 1997. Pavlov's physiology factory. *Isis* 88: 205–246.

Trout, J.D. 1994. A realistic look backward. *Studies in the History and Philosophy of cience* 25, no. 1: 37–64.

Trout, J.D. 1998. *Measuring the intentional world: Realism, naturalism, and quantitative methods in the behavioral sciences.* New York: Oxford University Press.

Trout, J.D. 2002. Scientific explanation and the sense of understanding. *Philosophy of Science* 69: 212–233.

Trout, J.D. 2007. The psychology of scientific explanation. *Philosophy Compass* 2, no. 3: 564–591.

Trout, J.D. 2008. Seduction without cause: Uncovering explanatory neurophilia. *Trends in Cognitive Sciences* 12: 281–282.

Trout, J.D. 2009. *The empathy gap: Building bridges from the good life to the good society.* New York: Viking/Penguin.

Turner, H. 1995. *Science in medieval Islam.* Austin: University of Texas Press.

University of California, San Francisco Medical Center. Biofeedback for Iincontinence. Retrieved from http://www.ucsfhealth.org/education/biofeedback_for_incontinence/#2.

VanFraassen, B.C. 1980. *The scientific image.* Oxford: Clarendon Press.

———. 1989. *Laws and symmetry.* New York: Oxford University Press.

Van Helmont, J.B. 1650. *A ternary of paradoxes.* Trans. W. Charleton. London. Printed by James Flesher for William Lee.

Van Swol, L., and J. Sniezek 2005. Factors affecting the acceptance of expert advice. *British Journal of Social Psychology* 44, no 3: 443–461.

Von Euxkull, J. 1934/1957. A stroll through the worlds of animals and men: A picture book of invisible worlds. In C.H. Schiller (ed. and trans.), *Instinctive behavior: The development of a modern concept.* New York: International Universities Press, pp. 5–80.

Vonnegut, K. 1963/2010. *Cat's cradle.* New York: Dial Press Trade Paperback.

Watson, J.D. 1968/1999. *The double helix.* London: Penguin.

Weinberg, S. 1994. *Dreams of a final theory.* New York: Random House.

———. 2001. Can science explain everything? Anything? *New York Review of Books* 48, no. 9: 47–50.

Weisberg, D.S., F.C. Keil, J. Goodstein, and E. Rawson, and J. Gray. 2008. The seductive allure of neuroscience explanations. *Journal of Cognitive Neuroscience* 20: 470–477.

Whitfield, T.W.A., and P.E. Slatter. 1979. The effects of categorization and prototypicality on aesthetic choice in a furniture selection task. *British Journal of Psychology* 70: 65–75.

Whittlesea, B.W.A., and J.P. Leboe. 2003. Two fluency heuristics (and how to tell them apart). *Journal of Memory and Language* 49: 62–79.

Winkielman, P., N. Schwarz, T. Fazendeiro, and R. Reber. 2003. The hedonic marking of processing fluency: Implications for evaluative judgment. In J. Musch and K.C. Klauer (eds.), *The psychology of evaluation: Affective processes in cognition and emotion.* Mahwah, NJ: Lawrence Erlbaum Associates, pp. 189–217.

Williams, G. 2010. *Angel of death: The story of smallpox.* Basingstoke: Palgrave Macmillan.

Wilson, A. 2008. Hydraulic engineering and water supply. In J.P. Oleson (ed.), *Handbook of engineering and technology in the classical world.* New York: Oxford University Press, pp. 285–318.

Woolf, V. 1979. *The letters of Virginia Woolf: 1932-1935*, vol. 5. Ed. N. Nicolson and J. Trautmann Banks. New York: Harcourt Brace Jovanovich.

Yates, J.F, P. Price, J. Lee, and J. Ramirez. 1996. Good probabilistic forecasters: The "consumer's" perspective. *International Journal of Forecasting* 12: 41–56.

Zarnoth, P., and J. Sniezek. 1997. The social influence of confidence in group decision making. *Journal of Experimental Social Pscyhology* 33, no. 4: 345–366.

INDEX